Making a Scientific Case for Conscious Agency and Free Will

Making a Scientific Case for Conscious Agency and Free Will

W. R. Klemm

AMSTERDAM • BOSTON • HEIDELBERG • LONDON
NEW YORK • OXFORD • PARIS • SAN DIEGO
SAN FRANCISCO • SINGAPORE • SYDNEY • TOKYO

Academic Press is an imprint of Elsevier

Academic Press is an imprint of Elsevier
125, London Wall, EC2Y 5AS.
525 B Street, Suite 1800, San Diego, CA 92101-4495, USA
50 Hampshire Street, 5th Floor, Cambridge, MA 02139, USA
The Boulevard, Langford Lane, Kidlington, Oxford OX5 1GB, UK

ISBN: 978-0-12-805153-5

British Library Cataloguing-in-Publication Data
A catalogue record for this book is available from the British Library

Library of Congress Cataloging-in-Publication Data
A catalog record for this book is available from the Library of Congress

For Information on all Academic Press publications
visit our website at http://store.elsevier.com/

Working together
to grow libraries in
developing countries

www.elsevier.com • www.bookaid.org

CONTENTS

All humans have the feeling that they consciously will certain things to happen and that in the absence of external restrictions they are free to choose among alternatives. This common experience makes us think we have free will. Yet in the last 35 years or so, scientists have been conducting experiments that they interpret as evidence that free will is an illusion. Their idea is that everything is driven by an unconscious mind that informs the conscious mind of the choices and decisions made. Since free will would require consciousness agency, antifree will advocates further claim a supporting corollary that consciousness can only "observe" and cannot do anything. In addition to the problem of trying to prove negatives, these experiments have major design and interpretation flaws that have been identified by numerous scientists, as I summarize in the chapter "The Scientific Case Against Conscious Agency and Free Will."

No coherent scientific argument has been made to support free will. Typically, people want to defend free will on grounds of social, legal, or religious perspectives. These arguments, appealing as they are, seem specious because they appeal to consequences not causes of choices and decisions (see chapter: Philosophical, Religious, Social, and Legal Arguments). A series of arguments are made later on human behaviors that are hard to explain without conscious agency and free will (see chapter: Physiology of Mental States and Conscious Agency). The treatise concludes with an examination of how conscious agency and free will can emerge from the materialistic processes of brain function (see chapters: Free-Will-Dependent Human Thought and Behaviors and Neuroscience May Rescue Free Will from Its Illusory Status). The best route for settling the debate lies in the neuroscience of the future, and only then if better tools for system-level analysis become available.

ACKNOWLEDGMENTS

Early drafts of this manuscript were critiqued by psychologist, Marc Whittmann, at the Institut für Grenzgebiete der Psychologie und Psychohygiene e.V. in Freiburg, Germany and neuroscientist Robert Vertes at the Center for Complex Systems and Brain Sciences at Florida Atlanta University. The author wishes to thank them for their specific contributions and, in general, thanks all those scholars who have dared to publish in this controversial area—irrespective of their conclusions.

CHAPTER 1

The Scientific Case Against Conscious Agency and Free Will

1.1 ILLUSORY FREE WILL

Scientific and philosophical fashion these days claims that humans have no free will. That is, we are basically biological robots, driven to our thoughts, beliefs, choices, intentions, and actions by unconscious forces in our brain. We are puppets controlled by the programming from our genes and life experience. Free will is deemed an illusion. Freudian psychology has been reborn in a revised framework of a pre-eminent unconscious mind (Fig. 1.1).

Free-will deniers get their idea by extension of the fact that the brain does make unconscious choices. If you accidentally touch a hot stove, it is your unconscious mind that initiates the hand and arm withdrawal reflex. Only afterwards are you informed, "Damn that hurts." Even at more complex levels of behavior, humans commonly make unconscious choices, as when our behaviors are stereotyped or compulsive. Certain disease states, such as obsessive/compulsive disorders, are driven unconsciously with little or no conscious control (that's why it is called a disease). Damage to certain areas of the brain's source of consciousness, the neocortex, can change personality-driven choices. The problem is that such phenomena are inappropriately extended to mentally normal people and the assumption that conscious does nothing but observe, that it has no agency.

The word agency has many meanings, but the meaning used here is the issue of whether human consciousness has the power to generate self-directed causes of thoughts, feelings, attitudes, beliefs, and behaviors—and associated choices and decisions. A recent book explores a wide range of theory about agency (Gruber et al., 2015) and I have a chapter there on the neurobiology of agency. All neuroscientists agree that the unconscious mind, when a person is awake, has agency in that it directs reflexes and more complex behavioral responses that do not require conscious intervention. But neither scientists nor philosophers think much

Making a Scientific Case for Conscious Agency and Free Will. DOI: http://dx.doi.org/10.1016/B978-0-12-805153-5.00001-8

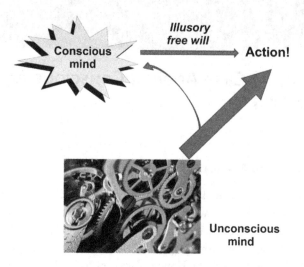

Figure 1.1 Diagram of the modern notion of illusory free will. Unconscious mind generates willed action and informs conscious mind, which acts only as an observer and has no capacity for agency.

about what the unconscious mind is doing during sleep and what it is about wakefulness that gives agency to the unconscious mind. The crux of the free-will debate focuses on whether any portion of a person's agency comes from conscious direction. I make a case for conscious agency in chapter "Physiology of Mental States and Conscious Agency."

The denial of free will has a centuries-old history, but is now popularized by an influential clutch of activist scientists, such as Richard Dawkins, Sam Harris, and Daniel Wegner, who have become intellectual rock stars from their best-seller books arguing the case against free will. Many philosophers also have joined the illusory free-will crowd.

Free-will deniers seem to believe there can be no ego, no real "I" that we all erroneously think we are. Descartes' "I think, therefore I am," can't be right because there is no "I." We are all "its." When we think we are making a decision or a plan or directing an action, it is not our "I" that does this. To buttress their position, many deniers have constructed the argument as follows:

1. Freely chosen actions would have to come from a conscious brain.
2. Consciousness is only an observer that has no agency.
3. Therefore, there can be no free will.

If consciousness is viewed only as an observer of a decision or choice made by the unconscious mind, then the conscious brain must not be the source of actions that occur during consciousness. This book tackles the subject of free will with trepidation. The subject, as Christof Koch (2012) puts it, "is a scholarly minefield." An obvious first step to counter the illusory free-will argument is to challenge its premise that consciousness lacks agency, which this book does in chapter "Physiology of Mental States and Conscious Agency."

1.2 KEY DEFINITIONS: FREE WILL, CONSCIOUS AGENCY

What do we mean by "free?" Apparently there is no clear consensus. One thing we can say with certainty: the brain's network of networks surely has more degrees of freedom than does a single neuron. The latest book on the subject by Joaquín Fuster (2013), *The Neuroscience of Freedom and Dignity*, makes a useful distinction between freedom of action and free will. I can illustrate Fuster's thesis this way: A mosquito is free to fly, but is unlikely to freely will to fly because flying insects lack the cortical network structure of humans. But humans are free to fly—by consciously inventing and building balloons, planes, and rockets. Fuster thinks that humans have an enormous amount of freedom to make choices, but they cannot freely will anything because such factors as genes and environment necessarily dictate brain function.

Before a case can be made for or against free will, we will need an appropriate operational definition. Such definition is not self-evident. Most scholars could probably agree on what free will is not: a decision without cause. Many decisions are forced by circumstance, but humans do make decisions that are caused by internal drives and preferences. We could not necessarily predict an action from knowing the DNA sequence, brain developmental history, and the programming from life-time experience. Nor could most actions be totally free from the factors just mentioned. Nor could free will arise from strictly random processes. The operational definition in this book is as follows:

Free will occurs when a person makes a conscious choice from multiple options, none of which are predetermined or compelled.

Intentions, decisions, and choices are of course influenced by their unconscious antecedents, but are not inevitably determined by them because the conscious mind can intervene, veto, or otherwise control. A simple parlor trick illustrates in a simplistic way a difference between unconsciousness and conscious influences. Clasp your hands together in front of your face with fingers overlapping. Raise each index finger and move about one inch apart. Shut your eyes. Amazingly, the fingers drift together, even though you did not consciously issue such an instruction. This is your unconscious mind exerting its will. (Actually, the movement is a spinal-cord-mediated reflex mediated through the brachial plexus that did not involve the brain.) Now repeat the process with eyes open but willfully try to prevent the drifting. With some mental effort, you can stop the movement. This is the "I" of your conscious mind freely exerting executive control.

We can also debate the definition of consciousness. It can be most simply defined as the opposite of unconsciousness, which is not having such agency of unconscious mind as reflexes and more complex automated behaviors. In this book, I will use as an expedient a colloquial understanding of consciousness. Consciousness has two elements, one of which is qualia (that is, the redness of red). More useful for the issue of free will debate is the awareness aspect of consciousness that exists explicitly in sensory perception, reason, anticipation, and planning for the future, and the making of choices and decisions.

Descartes idea of "I" was chided by critics as requiring a "ghost in the machine." My position, outlined in the chapters 4 and 5, is there is an "I" that is not a ghost but rather a materialistic process based on a constellation of special nerve impulse signaling patterns in the brain. Moreover, I suggest a way that consciousness can use the brain's neural machinery to exert executive control.

Even deniers concede that there are executive control networks in certain cerebral cortex areas, and these are unique for each individual partly as a result of the programming produced by a person's lifetime of experiences, choices, and decisions. Efficient executive networks must have some freedom to decide, plan, and act, even if they are constrained by genetics, contingencies, and interactions with other parts of the brain. Otherwise, the networks should not be considered as "executive."

1.3 DETERMINISM AND QUANTUM MECHANICS

If free will is defined in the context of physical determinism, then it cannot exist. Determinists may say that willed action that has a cause cannot be free. I regard that as a specious argument, because the possibility of free will is excluded by definition, not evidence or logic.

Determinism holds that a given process causes and predicts a consistent result. But can this principle be applied to brain choices and decisions? Many physicists claim there can be no free will because brains are controlled by physics, and physics allows no free will. Paradoxically, physicists show that quantum physics has uncertainty. If the location of electrons is not predetermined, why can't human thought be that way too?

The uncertainty principle of modern physics seems to undermine determinism. However, it may not be appropriate to invoke the same doctrine at all levels of organization, ranging from inanimate physics to biosystems to the human mind. While brain events are clearly caused, specific outcomes from the same causal input are not inevitable. Even in the face of constraints, a human can choose to do something else besides the favored choice generated by strong biological or situational forces. Thus it does not seem tenable to assert that determinism precludes free will.

Determinism can still be consistent with free will in that a freely made decision can cause the resulting action. The making of the decision also has a cause, as documented by abundant neurophysiological evidence (see section "How the Brain Makes Choices/Decisions" in chapter "Neuroscience May Rescue Free Will From Its Illusory Status").

The causation models of physics are typically reductionistic, bottom-up. Moreover, the essential determinism of Newtonian physics breaks down in light of later discoveries of the unpredictabilities of quantum phenomena at subatomic levels and chaos dynamics at system levels.

Some physicists presume that quantum mechanics (QM) relates to brain function. Many physicists argue that the brain is a living quantum computer (McFadden & Al-Khalili, 2015). One line of argument argues that quantum consciousness actually provides evidence for free will (Hammeroff & Woolf, 2003).

Quantum phenomena apparently operate in such living processes as photosynthesis, enzyme biochemistry, and cytochrome energy transport and capture. Some scholars suggest that life is defined as a balance between quantum and classical physics phenomena (McFadden & Al-Khalili, 2015). But nonliving matter also has a balance between the two levels of physics. Life remains mysteriously different.

Quantum explanations exist for dynamical change in protein conformation, and these have been used to develop QM theories of consciousness (Hammeroff & Woolf, 2003). Subtle conformational changes of protein in ion channels and second messengers are a proximate cause of nerve impulse generation. But what is serving as signal for information processing and messaging? Most likely it is the propagating steam of impulses in a spike train (Klemm, 2011c), which can be explained by the nonlinear dynamics of traditional physics. Quantum phenomena are not evident at this level.

Nonetheless, QM might have a profound indirect influence on consciousness. Hammeroff and Woolf argue that the collapse of quantum superposition is neither random nor deterministic and thus could account for free will. Also, they note the vast computational capacity provided by the enormous number of bit states of cellular proteins in neurons and their synapses, an operation with some 10^3 operations/second. But there is no evidence yet that this numerical computation capacity determines consciousness. In fact, in so-called split-brain subjects where the two hemispheres are surgically isolated, each hemisphere is conscious with only half the neural machinery. Each hemisphere, however, is conscious of fewer items than is the whole brain.

Physicists prove that light photons are "entangled," that is, their behavior in one place can be linked to behavior of photons in other far-distant places. The problem for brain theorists is that no evidence supports the idea that light has signaling properties inside the brain (in fact, it is dark in there).

Likewise, entanglement and "tunneling" of electrons can be demonstrated in certain physical systems, but electrons do not convey or process information in the brain. Positively charged ions that have lost electrons account for neural signaling. And the processing occurs in macromolecular synaptic neurotransmitter and second messenger systems.

Physicists show that subatomic particles have "superposition," that is, they can be in two different states at the same time. So ...? What has that got to do with neural information processing and messaging?

Scott (2003) questions QM's relevance on the grounds that brain function is nonlinear, whereas QM is linear, as expressed in Schroedinger's equation. Nonlinear functions can yield emergent properties that are greater than the sum of their parts. Maybe it is the nonlinear neural processes that create the emergent property of consciousness.

In sum, there is no evidence that the brain is a quantum computer. Though QM seems unpromising as a way to explain consciousness, we must concede that classical physics has not explained it either. A major purpose of this book is to suggest the kinds of research on classical nonlinear neural dynamics that might lead us closer to understanding consciousness.

Moreover, the history of science has demonstrated that knowledge and understanding of reality can be illusory and transient. Unpredictability need not imply randomness, but rather could "allow the metaphysical possibility that there are further causal principles at work in bringing about the future beyond those that are described by science's bottom-up notion ..." (Polkinghorne, 2009, p. 84). In other words, as Polkinghorne elegantly put it, "Science trawls experience with a coarse-grained net, and much that is of the greatest significance slips through its wide meshes" (p. 93).

> *The determinism of physics does not inevitably apply to the electrodynamics that exists in the activity of neural circuits that compete with each other to make a selection among nonmandatory multiple options.*

1.4 MENTAL STATES ARE PHYSICAL STATES

The sense of self is the mental state most crucial to our perception of free will. It is "I" who thinks, feels, decides, etc. Some critics of free will seem to assert that willed action is caused by the brain and that therefore there can be no free will exerted by the self. But what is the self? Is it not brain function? As I will elaborate in chapter "Physiology of Mental States and Conscious Agency," we cannot usefully address the issue of free will until we resolve the questions about the neural basis of self and whether conscious self has a capacity for agency.

What is meant by "mental state?" A major source of confusion about free will arises because laymen and scholars alike commonly make a distinction between neural and mental states. On what grounds can we assert that mental state is somehow separable from brain function? "Mental" is simply a metaphor for human brain function. Later in this book I will explain my view of "mental" in materialistic terms of brain function. Considering mental and physical states as synonymous avoids the necessity for saying that mental states are incapable of supervening brain function to exert free will.

Mental states are not epiphenomena; they are neural states. This common distinction originated centuries ago when enlightenment from neuroscience was not available. Mental states are physical, neurophysiological, states with a capacity for agency because they are neurophysiological. Lower-level unconscious causes do not rule out higher-level conscious causes. Indeed, they may even operate conjointly.

Mental operations, because they are of materialistic origin, are capable of some top-down control of bottom-up processes. The determinism of physics does not inevitably apply to the electrodynamics that exist in the activity of neural circuits that compete with each other to make a selection among nonmandatory multiple options. See chapter 5 for elaboration.

Another semantic trick used to reject free will imputes an immaterial soul as a prerequisite. But free will does not depend on an immaterial soul. Brain processes can come freely from internal self-organizing capabilities of complex networks, as I explain in chapter 5.

The serious challenge of free will comes from assumptions that the conscious mind has no agency, is only aware of choices and decisions, without the ability to make or alter them. Searle (2007) argues the point that consciousness helps the brain freely arrive at decisions. Conscious thinking uses motives and reasons as a means to an end. Consciousness creates an intention to arrive at an appropriate decision, free in the sense that the final decision is neither preknown nor inevitable.

1.5 THE ORIGINAL FREE-WILL RESEARCH

In terms of biological science, research by Benjamin Libet and his followers that began in the 1980s was the major source for the current free-will debate. Many investigators interpret a series of related tests of free will to indicate that certain brain activity increases not only before a specific willed movement, as expected, but also a fraction of a second *prior to* the conscious realization that the decision to move was made. The untested assumption was that this early activity was solely due to unconscious decision-making.

These studies generally used a similar concept and design, which have been roundly criticized on objective criteria in publications by more than a dozen other scientists (eg, see Baker, Mattingley, Chambers, & Cunnington, 2011; Grill-Spector & Kanwisher, 2005; Guggisberg & Mottaz, 2013; Haggard & Eimer, 1999; Herrmann et al., 2008; Jo et al., 2014a, 2014b; Klein, 2002; Klemm, 2010; Lages & Jaworska, 2012; Obhi & Haggard, 2004; Radder & Meynen, 2012; Restak, 2012; Schlosser, 2014; Schurger et al., 2012; Tempia, 2011; Trevena & Miller, 2010).

Flaws in the research fall into several categories: (1) premise deficiencies, (2) technical limitations in experimental design, (3) misinterpretation of events preceding the decision, (4) unreliability of self-reported decision, and (5) overdrawn generalizations of the implications. By category, the flaws include the following.

1.5.1 Premise Deficiencies
- Separation of our subconscious and conscious minds. Both exist simultaneously and interactively in the same brain.
- All-or-none thinking. You either have free will or you do not, and the possibility of partial free will is not accommodated.
- Experimental designs based on an inappropriate definition of free will or that are testing something other than free will.
- Behaviors, such as button pressing, that are too simple and automated to be representative of most real-life decisions.
- The unproven assumption that a decision is a point-process that can be reported with millisecond resolution.
- Failure to consider that reporting of when a conscious thought occurred can be delayed from when it actually occurred.

1.5.2 Technical Limitations in the Design
- Statistical deficiencies, especially in the fMRI studies.
- Uncontrolled variables in many of the experiments.
- Experimental designs that did not provide opportunity for reasoned analysis or planning.
- Averaged electrical signatures of "decision" data in most studies, without tracking of individual trials (which did not always show the purported effect).
- Limited use of different test designs that minimize the deficiencies of the prevailing paradigm.

1.5.3 Misinterpretation of Antecedent Events

- Failure to explain why some individual trials produce an opposite electrical signature of decision-making.
- Assumption that all antecedent events reflect sequential processes. Some can occur simultaneously in parallel and some may be conflated with the actual decision-making.
- Some antecedent events, especially those that have been recorded in nonmotor brain areas, are an inextricable part of the conscious component of decision-making processes, such as urges, intentions, and plans.
- Failure to consider the possibility that a mixture of unconscious and conscious processes generates a given decision, each with their own antecedents and time relationships to the action.
- Inability to measure the relevant conscious introspection and reasoning and especially its time relationship to the "moment" of decision.

1.5.4 Unreliability of Self Reports

- Assumption that humans can accurately identify on a millisecond timescale when they have a specific thought.
- Unrecognized illusions in the mental perception of time and sensation.
- Short timeframes, often in milliseconds. Even simple decisions can spread out over several or more seconds.
- Unwarranted assumptions about how long it takes to make and then to realize that a conscious decision has been made.

1.5.5 Overdrawn Generalizations of the Implications

- Unwarranted extrapolation of results of simple, reflex-like responses in a laboratory setting to the more complex decision-making situations of real life.
- Untested rejection of conflicting data and interpretations.

In the interests of completeness, I should mention the work by Wegner (2005) who provides a different kind of evidence for his claim of illusory free will. He showed that under special test conditions a person can wrongly assume they have willed a certain action, but this by no means proves that this is the case under all conditions, especially those that occur outside the laboratory.

The author of the most recent book on free will (Fuster, 2013) was either unaware of the critical published research or willfully chose to ignore it. Fuster claims that "99 percent of all actions will be unconscious," and therefore devoid of any free-will source. This strikes me as unfounded, and the 99% statement is typical of the sloppy thinking in this field. As an example of the common prejudice against free will, a leading neurophilosopher, Patricia Churchland, provides a book cover endorsement that calls Fuster's work a "masterful accomplishment."

Fuster and his sympathizers uncritically endorse Libet-style research and conclude that consciousness can only participate in goal-setting, reasoning, and planning, while intention, choice, and action are strictly unconscious processes. Is it not strange that he fails to consider that goal-setting, reasoning, and planning are inextricably bound to decision-making?

Paradoxically, Fuster gives humans more social and political freedom, while denying that they have any freedom in deciding how to act. For example, he says that people have freedom because genetics and life experience create their cortical complexity. Specifically, "The individual with a richly interconnected cortex, intelligent, well-schooled, and with superior linguistic skills will have more options in life, and thus will be in principle more free," than individuals without these benefits. He refuses to concede that a person could freely choose some of the options that would affect schooling, linguistic skills, and even intelligence.

Most recently a report has appeared that free-will critics may glom on to, a fMRI study that showed that connectivity patterns were unique to each individual, personalized much like fingerprints (Finn et al., 2015). The connectivity profile is intrinsic and independent of task and rest conditions. But evidence for being unique is hardly relevant to the issue of free will. First, the connectivity study shows that it is the metabolism that is linked between brain areas and that reveals nothing about linkage or permanence of neural signaling. Secondly, one could argue with equal force that people can have unique capabilities for free will, some more, some less. In fact, casual observation reveals enormous individual differences in mental stereotypy and compulsivity, as opposed to flexibility, and creativity.

The free-will deniers' hypothesis of conscious agency and free will cannot meet the standard test of hypotheses: experiments should be able to falsify a wrong hypothesis. To say there is no free will is a negative, and you can't prove a negative. Moreover, all evidence indicates that conscious and unconscious minds in the awake state are inextricably linked and without clear demarcation.

So, is there a reasoned counter-argument? Eddy Nahmias (2011) thinks that the illusory free-will position is "too quick and glib." He says "It is like inferring from discoveries in organic chemistry that life is an illusion just because living organisms are made up of non-living stuff."

The ideology of illusory free will assaults morality, the law, religious belief, and common sense. Nonetheless, assaulting common sense seems to delight many elitist scientists who find gratification in thinking that their sense rises above that of commoners and lesser intellects who think they have free will.

I think the illusory free-will arguments have been dismantled by the refutation of the experimental evidence. As long as Libet-type experiments can't pass analytic muster, free-will illusionists have no real evidence for their assertion that all intentions are generated unconsciously and are inaccessible to consciousness until after the moment of decision. The importance of discrediting the supposed evidence for illusory free will is the credence it gives to the only alternative that there can be at least a degree of free will. I intend here to make a case that humans do have some free will, existing at a significant level even if not complete.

CHAPTER 2

Philosophical, Religious, Social, and Legal Arguments

A long history of debates about free will can be found in the scholarly literature. Lay people, and some scholars, sometimes defend free will on the grounds of certain religious, social, or legal perspectives. Their arguments are based on the consequences of not having free will and provide no direct evidence for free will.

2.1 PHILOSOPHICAL ARGUMENTS FOR FREE WILL

When it comes to philosophical arguments about free will, there are only three choices. You either believe it exists, does not exist, or partially exists in conjunction with nonfree determinants. To many philosophers, free will's existence is a wide-open question (Balaguer, 2010). Philosophers generally are all over the map on this issue. On the fringe are those who share the Buddhist view that there is no "self" that can initiate any kind of will, free or not. Philosophers who advocate for free will do so in the limited "compatibilist" sense that humans generally lack free will but can sometimes have free will.

Philosophers who reject free will are often called "determinists," because they believe that every choice or decision is caused by prior events. Of course, every mental choice or decision has a neural cause, but whether a given choice is inevitable is the contested issue. When determinist philosophers buttress their arguments with science, it is typically the science of physics; namely, all physical events have physical causes. They assume that a contrary view would invoke the idea of a nonmaterial "mind." I hope to show in later chapters that "mind" is material and can also exert a degree of freedom. Determinist philosophers also rely on the flawed neuroscience in the 1980s, as summarized in the chapter "The Scientific Case Against Conscious Agency and Free Will."

Making a Scientific Case for Conscious Agency and Free Will. DOI: http://dx.doi.org/10.1016/B978-0-12-805153-5.00002-X

Philosophy can support both sides of the free-will argument, even both at the same time, as in compatibilist philosophy. Compatibilist philosophers leave some room for free will. Their less rigid position was apparently introduced centuries ago by Thomas Hobbes. Prominent contemporary leaders in free-will philosophy like Daniel Dennett (2014) and Alfred Mele (2014) take issue with the illusory free-will idea on the grounds that harmonious social and moral behavior would not be possible without free will. Dennett calls the illusory free will conclusion a "morally pernicious idea."

Compatibilists believe that the unconscious mind drives much of what humans believe, think, and do, but that consciousness has an agency that is not compelled to follow unconscious demands. Though we owe gratitude to philosophers for championing the debate, philosophy is not science. Philosophers are prone to play obscurant word games about free will (Tye, 2009). Tye argues that consciousness has to have a physical cause, but strangely argues for mental causation in a material world, as if mind itself were not material. He speaks of "phenomenal externalism," which confuses me. But I am reasonably certain this this view, whatever it means, does little to make a case for conscious agency or free will. Conscious mind and free will need to be considered in neuroscientific terms which neither Tye nor others do very well. This present book does make that attempt.

The key to understanding is to realize that a conscious brain is aware of and experiences life in the context of a conscious self-awareness. Since free will, by definition, would require consciousness, the issue becomes one of establishing that consciousness has agency. Then, if consciousness has agency, the issue becomes one of establishing a degree of freedom for that agency. All of this can be a material process, as I suggest in the chapters that follow.

2.2 RELIGIOUS BASIS OF FREE WILL

Religion provides a main source of free-will doctrine. Ideas of moral responsibility originate in religious views of right and wrong and the belief that followers have the capacity to make the correct choices. Their free-will capacity makes them morally accountable.

The materialistic view of consciously mediated free will that I will present throughout this book does not address the primary essence of all religions: an immaterial "soul" or "spirit" that survives death of brain and consciousness. Free will does not have to involve souls or some dualist conception, as I try to explain in chapter 5. Free will can be mediated by materialistic processes that we already mostly understand and know how to study. Scientists tend to insist that free will requires dualism, and in the process make free will a strawman argument they can easily knock down. Today's science has no way to test for a soul. But, we have no obligatory need to assume that free will requires an immaterial soul.

Religious believers will have a hard time accepting the view that personhood resides in neural networks. This does not mean that "soul" does not exist, but soul, whatever it is, cannot be neural networks as we understand them. We have no idea how an immaterial soul could be coupled to these interacting neural networks nor how a materialistic mind in this world can be replicated or carbon copied in another space–time dimension. That is an issue of faith, not science.

Nonetheless, we have every reason to believe that the universe contains material properties that we cannot describe or explain, such as the dark matter and dark energy that are vastly more abundant than the "material" things that we are still learning about. The existence of an "arrow of time" involving how "material" things exist in past, present, and future is not fully understood by physicists (Ellis, 2007). These are things we know we don't know. No doubt there are other phenomena that we don't know what we don't know. Thus, it seems intellectually imprudent to reject religious belief, as apparently most scientists do. Many of the prominent anti-free-will advocates also happen to be atheists. Not surprising. Otherwise, their position would be cognitively dissonant.

Many religions advance the argument that God has created humans who are free to choose to worship or reject him, and eternal reward and punishment depend on their choices. Religious doctrines that would deny free will must accommodate the logic that a judgmental god sadistically and unfairly punishes people who reject him, even though they had no freedom to make such a choice.

Christianity is uniquely individualistic and demands personal responsibility and accountability. Kierkegaard railed against growing collectivism, including that of organized religion. His emphasis was on the responsibility of individuals to be true to Christian ideals.

Yet religious doctrines are confused about free will. The Bible gives license for Calvinism's doctrine of predestination. The Qur'an explicitly rejects free will in many places but also stresses that God will judge people according to their deeds.

The point remains: the issue of free will remains whether or not one believes in God. Atheists make all sorts of choices and decisions just like believers. The question persists: are any of their choices freely made?

2.3 SOCIAL AND LEGAL BASES OF FREE WILL

Defenders of free will often make the correct argument that we must believe in free will or otherwise social consequences are meaningless. Mele (2014), for example, argues that humans not only have free will but a "deep" level of freedom to choose among many current options, based on data showing that bad behavior increases as people come to disbelieve in free will and personal responsibility. But such arguments are also flawed. Whether or not people believe they have free will surely affects their behavior, but that is not evidence for correctness of such a belief.

Accepting the illusory free-will world view would strip us of human capacity and personal responsibility. Clearly, it makes sense to live as if we have free will, believing in it or not, for otherwise we limit our chances for personal growth and fulfillment. We limit our capacity for free will by our willingness to claim and exert the free will we think we have.

Believing in free will has the enormously beneficial social, political, and economic consequences of promoting self-improvement, accountability, and cooperation with others (Rigoni & Brass, 2014). But such belief provides no evidence on whether or not free will actually exists. Religious, social, or legal arguments for free will are merely statements of supposed consequences of behavior without free will. I will propose a better defense.

A robotocist world is not only morally pernicious, but would be an affront to human dignity and personal development. People without free will are more likely to be victims and less able to change maladaptive attitudes and behaviors. Thus, it is argued that society and government must help people do what they cannot do for themselves. If we can't make conscious choices, then we can't do much to improve ourselves or our plight in life. We are just victims of genetics and circumstance. Or even if there are things that can be done to change us and our situations, the approach will surely have to be different if we can't initiate the change by force of our free will. Without free will, some outside force must program our unconscious.

We simply must hold people accountable for their actions, especially in the case of criminal and extreme antisocial behavior, like terrorism for example. But this is an argument about consequences of not having free will. The desired end of personal accountability does not justify a conclusion about free will one way or the other. At a personal level, people choose to avoid certain socially unacceptable behaviors because of the likely penalty. For example, you decide not to rob a bank because you are likely to go to prison. That doesn't prove you have free will. It may only mean you have learned about such scenarios and that learning may operate unconsciously.

> *Religious, social, or legal arguments for free will are based on the consequences of lack of free will. That does not provide evidence for free will.*

Similar analysis applies to legal issues. Legal analyst Jeffrey Rosen wrote in *The New York Times Magazine*, "Since all behavior is caused by our brains, wouldn't this mean all behavior could potentially be excused? ... The death of free will, or its exposure as a convenient illusion, some worry, could wreak havoc on our sense of moral and legal responsibility."

People who conclude that no-one has free will must then hold no-one responsible for criminal or evil acts. Free-will deniers would not make much effort to change and improve themselves. They could become intellectually and emotionally paralyzed. But again, the argument is about consequences of a belief, not whether or not there is evidence for the correctness of the belief.

2.4 HOW SCIENCE HAS OBSCURED THE ISSUE OF FREE WILL

Many scientists regard a scientific defense of free will as taboo. On the rare occasions when this happens, scientists tend to make specious arguments. For example, the famous neurosurgeon Wilder Penfield claimed that his experiments indicated the existence of free will (Penfield, 1978, p. 77). They do not. During surgery, he demonstrated that he could force people to make certain movements by electrically stimulating specific areas in the motor cortex. Invariably, the subjects said that they did not will the movements, that they must have been forced by the stimulation. Penfield described the results this way:

> When I have caused a conscious patient to move his hand by applying an electrode to the motor cortex of one hemisphere, I have often asked him about it. Invariably his response was: "I didn't do that. You did." When I caused him to vocalize, he said: "I didn't make that sound. You pulled it out of me". When I caused the record of the stream of consciousness to run again and so presented to him the record of his past experience, he marveled that he should be conscious of the past as well as of the present... He assumed at once that, somehow, the surgeon was responsible for the phenomenon... The electrode can present to the patient various crude sensations. It can cause him to turn head and eyes, or to move the limbs, or to vocalize and swallow. It may recall vivid re-experience of the past, or present to him an illusion that present experience is familiar, or that the things he sees are growing larger and coming near. But he remains aloof. He passes judgment on it all...If the electrode moves his right hand, he does not say, "I wanted to move it." He may, however, reach over with the left hand and oppose the action ... There is no one place in the cerebral cortex where electrical stimulation will cause a patient to believe or to decide.

Penfield not only assumed that people could freely will movements, but that in such cases the commands were coming from outside the brain. In addition to questioning his dualism, we should recognize that, of course, no one place in the brain generates a willed action. It is a system function, involving especially the primary components of the cortex's executive control system. Penfield could not discover this because he was stimulating through single electrodes restricted to single areas of cortex, not the multiple "upstream" areas that normally activate the motor cortex. He was stimulating the final command pathway for motion, not the areas of the brain that tell these final output neurons what to do. In his time, the executive cortical control networks had not been discovered. Even if Penfield had stimulated all the executive control areas with a battery of electrodes, he could not know in advance what stimulus parameters would be relevant, unless they had been determined by prior experiment. Situational context would surely influence the consequences of a battery of stimuli.

Most scholars think that science should be the final arbiter over free-will debates. The problem is that they typically pick the wrong science: physics, in particular quantum mechanics (QM). QM holds that subatomic particles, like electrons, have indeterminate positions. Without apparent cause other than their inherent energy, they randomly flit around the nucleus in unpredictable locations. Applying such phenomena to human thought is a category fallacy, because thought is a macrolevel process, not a microlevel one (Musser, 2015).

The problem with QM is the subatomic categorical level at which it operates. The random flitting of electrons is clearly indeterministic, but we should not assume that randomness scales with increasing the level of function to cells, brains, and minds. In fact, I have had professional statisticians tell me that hardly anything is random and independent in the real world in which humans live.

So then, is this an argument that mental intentions, choices, and decisions are deterministic? Well, maybe, but only in the sense that these processes have a cause. But this is irrelevant, because the real issue is whether any caused will can be freely made.

Physicists in particular like to argue that all information is physical and that physical matter has no emergent properties: therefore there can be no emergent free will. The problem is that the decisions expressed in willed action come from information processing that involves executive control over selection, predictive evaluation, and recombination of alternatives. It is the processing that yields emergent properties that distinguish thought from particle physics.

Intentions, choices, and decisions can be generated randomly, but human thought is usually not random. Processes do not have to be random to be indeterministic. Neural processing can yield varying results. The results of deliberative processes are not inevitable and do not always yield predictable results, especially when the thought is new or creative.

The universe at the macrolevel may be deterministic, but conscious human thought resides in its own self-organized universe in which information is processed in ways that are neither random nor predetermined. This capability of the brain to process information provides a way to generate free will.

If any physical science is relevant, it should be information theory (IT). Inasmuch as the higher-level function of consciousness is information processing, we should consider the standard IT, which is expressed by the equation,

$$\log I = \frac{1}{p}$$

> *What distinguishes thought from particle physics is that neural circuits process information to yield emergent properties.*

The amount of information, I, in a system can refer to the existence of events, such as the number or defined patterns of nerve impulses. The equation holds that the amount of information is inversely proportional to the probability (p) of event occurrence. Thus, unexpected or low-incidence events contain the most information. Events that are deterministic carry relatively less information than indeterministic ones. Thus, freely willed events have special informational value. We might expect a highly evolved brain to have this kind of information processing and messaging capability.

Notably, the information content of QM phenomena, like flitting electrons, is never processed. They just exist as is. By processed, I mean nonrandomly changed, in this case by other electrons or anything else. Brain function, however, takes IT to a new, nonrandom and indeterminate level by its processing of its information carrier, the propagating nerve impulse trains and algebraic changes in synapses of neural circuits.

However, it is a fair challenge to ask, "Computers process information. Why don't they have free will?" Here are my answers:

- Computers are not conscious of what they do.
- Computers process information digitally. Computations in the brain are analog.
- The number of brain processing computational elements far exceeds what is possible in computers, having 86 billion neurons, each with 100–1000 synaptic contacts, embedded in neural circuits that flexibly adjust their mutual functional connectivities. A given neuron may have multiple functions and be functionally recruited into

multiple circuitry. Brains can even change their "hardwiring" by up/downregulation of synapses and in some regions by birthing new neurons and circuit wiring.

- Computers have far less degrees of freedom and flexibility than brain and thus have less capacity for self-adjusting their processing operations.
- Unlike brains, computers cannot do anything they have not been programmed to do.
- A computer has limited ability to self-organize its processing algorithms.
- Computers have no consciousness and thus much less awareness of the feedback from their own processing and executive control ways to consider choice options and adjustments.
- Unlike the stereotyped processing of computers, brains do not process rigidly because brains have self-organizing capabilities that process information in ways that can be new and not predetermined.
- The pervasive nonlinearity of brains enables emergent properties that are greater than the sum of the parts.

CHAPTER 3

Physiology of Mental States and Conscious Agency

3.1 THE AUTONOMOUS SENSE OF SELF

The brain situates itself in its body. An unconscious sense of self resides in the topographical organization of how the brain maps the body, both for sensation and for movement. This may remind readers of the classical idea of "homunculus." But this mapping is much more than that—it registers self-hood. Sensations received from outside the body are detected as "nonself." The circuitry for self-identity allows the brain to know it has body parts, where they are, and what they are doing, both within the self and in relation to the nonself environment. The conscious sense of self knows where its body input comes from and knows when the input originates within or outside the self. The feedback from musculature is likewise critical for distinguishing self from nonself. The brain is made unconsciously aware of its body by the feedback from tension sensors in muscles and tendons and balance sensors in the inner ear. This mapped information allows the brain to create body schemas and images that track changes in position and movement.

Such a self is a unique being, sculpted over time by gene expression, experiences, and learning and memory (Klemm, 2011a). This same being has a periodic conscious state, in which the self and its environment are made explicit.

This mapping allows the brain to think, however unconsciously, about the relationships of brain to body and body to the external world. Throughout life a person's sense of self, conscious and unconscious, is sculpted by experience and learning. Choices, whether freely made or not, affect life experiences and add to the sculpting of self.

Making a Scientific Case for Conscious Agency and Free Will. DOI: http://dx.doi.org/10.1016/B978-0-12-805153-5.00003-1

The sense of self begins its topographical construction in the womb as the nervous system constructs sensory and motor maps of the body that permit the brain to know where sensory information is coming from and where to activate muscles for appropriate movements (Klemm, 2012a). But the body mapping is not just confined to sensory and motor cortex. Basal ganglia and the cerebellum participate in the implementation of body mapping, as of course do the numerous fiber pathways of these maps. In this way the brain forms bidirectional communication links with its body and in the process learns that it exists as an embodied self with its own identity.

As the fetal neocortex reaches its final development, such a brain not only "receives" information about itself but likely "perceives" its sense of self—consciously. Why do we think that late-stage fetuses have a capacity for conscious realization of self? By 32 weeks of human pregnancy, J.L. Conel's histological studies (1963) reveal that the neocortex, the seat of consciousness, is fully developed qualitatively in terms of neuronal cell types and the structural layering. At around the seventh month of gestation, human fetuses show bodily signs of dreaming (muscle twitching, rapid eye movements), which we know from the dream content of later life makes us aware of ourselves as observers and participants in the dream. Electroencephalograms (EEGs), which are not practical to record from human fetuses, show clear evidence of dream sleep from implanted electrodes at the beginning of the last trimester of sheep fetuses (Schwab et al., 2009). Dream states are conscious states in that we are aware of ourselves as autonomous agents of decision and action in the dream.

"Conscious mind" is a metaphor for the bioelectric processes of brain that create awareness that we are aware of certain things in the environment and our internal feelings and musings. The word "mind" has fogged our thinking about who we are. The word is a metaphor for our personhood. Metaphors have so much utility in scientific communication that scientists use them too casually. Metaphors are not substitutes for reality. The physical reality is that what we call mind is the set of electrochemical processes used by the brain to register sensory input, process and memorize it, and exert control over physiological function and behavior.

The reality of "mind" is that personhood resides in patterns of nerve impulses or the stored capacity for generating those patterns (Klemm, 2013). The real world does not exist as such in the brain. What we consciously perceive is:

Selectively Attended at the expense of much that is not attended,
Deconstructed (via receptive fields, feature extraction, transduction),
Represented in nerve impulse patterns,
Mapped in sensory cortex (by topographical maps of the body), and
Reconstructed (by binding of the deconstructed impulse patterns).

What we believe and think is:

Processed (by synaptic modifications in neural networks),
Related to Memory Stores (by working memory),
Reprocessed (by attending feedback from mental reflection and actions), and
Selectively Incorporated Into Memory.

Conscious sense of self is a functional state, but it is also a state of being. We see ourselves as agents, think we can be in control of our self, and at least sometimes are in fact in control of our self.

When the sense of self emerges consciously, it permits a degree of such processes as language, flexibility, reason, patience, will power, memorization, and creativity in ways not demonstrated unconsciously. When self emerges consciously, it operates as a reality check and aid for unconscious thinking. Conscious sense of self would thus differ from unconscious sense of self by having a different degree of autonomy and opportunity for flexible choice- and decision-making.

3.2 REFRAMING THE FREE-WILL ISSUE

Reframing the free-will issue needs to begin by understanding consciousness, because by definition free will has to be exercised consciously. Whenever decision and conscious agency are so inextricably coupled, the issue becomes a matter of what consciousness is and can do.

Scholars who assert that consciousness cannot do anything other than "observe" often do this because they are repelled by their idea of "mind" as dualistic. Free-will deniers bolster their position by claiming, without evidence, that consciousness can't do anything. Therefore, with a dismissive wave of the hand, they assert that consciousness cannot generate any freely willed action. But we can readily imagine that even in the simplistic Libet-type experiments, subjects consciously must think about such antecedent elements as: "What am I supposed to do? ... be in control so I can provide clean data to the researcher, keep my eye on the clock, pick a time to act (press a button) ... immediately note the clock hand position when I decide to act ... press the button ... remember the exact time long enough to report it to the experimenter... figure out when is a good time to repeat all these actions." Thus, free will, to the extent it may exist, could not only be a vital component of decision-making but also of many consciously mediated antecedent elements and their implementation.

Because free will has to be expressed in consciousness, we must consider the role of consciousness in general. Consciousness raises the bar for proving free will is an illusion because it provides the ability to select a focus of attention to stimuli to which the person wishes to respond. To what extent is such selective attention freely made? What experiment shows that selective attention is random or deterministic? The conscious mind selectively attends stimuli that are perceived to be potentially salient. Such perception involves the capacity for consciousness to "do" the four things outlined above (process, relate to memory, reprocess, selectively incorporate into memory).

To reject the notion of free will, there is a series of human behaviors that need to be explained away if there is no such thing as free will. The series includes unpredictability, patience, reason, will power, deliberate memorization strategies and tactics, and creativity—as will be subsequently explained.

Finally, recognizing that willed action, free or not, is generated in neural circuits, we must explain how such circuits make choice and decisions. Specifically, we should seek to understand how neural circuits weigh options and make final selections. In the process, we may see what a degree of free choice could mean in materialistic terms.

3.3 NATURAL SELECTION AND EVOLUTION

If human consciousness is the product of evolution, why would it not have agency? Natural selection could be expected to favor evolution of both conscious agency and free will. Both are useful in the complex decisions that humans are capable of making, in such forms as:

- Explicit awareness of situational contingencies.
- Identification of the targets of selective attention.
- Rational analysis, often involving language, of appropriate intentions to respond or initiate action.
- Awareness of available options and rational analysis of the pros and cons.
- Anticipation of intended and perhaps unintended consequences.

Each of these requirements would benefit from having a capacity for conscious agency and free will. Ask yourself this: What advantage does human free will have over being a biologically programmed robot? Framing free-will issues this way makes an evolutionary argument highly unavoidable.

What purpose is served by evolving consciousness if it can't do anything? Natural selection forces usually favor traits and capacities that serve some useful purpose. Consciousness therefore is likely to be useful. This applies also for free will. With free will a person is much more likely to explore and seek new and adaptive experiences. Even just believing in free will does that. Free will offers an escape from the constraints of genes and experiential programming, allowing us to ask a wider array of questions and think more expansively. Attitudes and behaviors can be more flexible and creative. The conscious mind seems to be able to veto or modify an ongoing action "on the fly." Veto capability is even acknowledged by many critics of free will.

At a minimum, consciousness is an indirect cause of decisions in that consciousness can teach the unconscious mind what learning situations to engage, which to remember, and what the brain needs to think and decide. As mentioned, the conscious mind selectively attends much of what it experiences, and what is selected is often not obligatory. The conscious mind is thus personally responsible for its own programming. People tend to deny responsibility when poor choices are made. Excuses are psychologically convenient, but rationally hard to defend.

The felt need for excuses is itself an indication that the conscious mind knows it had the free will to have done better.

Such consequences of free will are certainly advantageous in generating a progressive culture in which the human species can thrive. Adventure-seeking and creativity open new vistas for humans, individually and collectively. Thus, forces of nature would select for preserving the genomes of people who are adventure-seeking and creative. Seeking adventures into the unknown surely led to the rapid expansion of humans to move out of Africa and spread around the world. Creativity led and still leads to innovations in agriculture, engineering, medicine, and culture that benefit the human species. It could well be that consciousness had a natural selection advantage because it provides the brain a way to have a capacity for free will.

> *The felt need for excuses is itself an indication that the conscious mind knows it had the free will to have done better.*

3.4 THE "MERE OBSERVER" ARGUMENT

Consciousness certainly "does things" in the sense that what we think affects neural processing in various parts of the brain. Multiple brain-scan studies indicate that conscious forces have specific influences on networks in specific brain regions. Moreover, placebo effects show that expectations and beliefs markedly modulate activity in brain areas responsible for perception, movement, pain, and various aspects of emotion (Beauregard, 2007).

For free will to be expressed, the conscious mind must be able to do things and not simply observe what is served up from unconscious processes. So the essential position of free-will deniers is that they must also deny that consciousness can direct our thoughts and actions—many deniers make such a denial. A popular view is that consciousness is only an "observer" and "interpreter" (Gazzaniga, 1998). Conscious executive control, intuitive as this notion may seem, is challenged by a significant number of scholars (Di Pisapia, 2013; Haggard, Clark, & Kalogeras, 2002; Pockett et al., 2006; Wegner, 2002, 2005). This myopic view of consciousness has little supporting evidence and many contrary arguments and evidence.

Most damaging to the mere observer argument would be the requirement for the unconscious mind to do everything a conscious brain does. As explained in chapter "Free-Will-Dependent Human Thought and Behaviors," many brain functions and behaviors cannot occur during nondreaming sleep. If consciousness is necessary to unleash unconscious capabilities that cannot be expressed in sleep, then where is the evidence that this is the case and how might it be done? No matter, this alone would be prima facie proof that consciousness has agency and is more than an "observer."

3.5 CONSCIOUSNESS AS A MEDIATOR OF AGENCY

A conscious sense of self yields the sense of personhood explicitly. It is "I" who believes, thinks, feels, decides, and acts. The conscious sense of self is certainly not an illusion, inasmuch as self-hood has physical representation in the unconscious mind. What is the "I" in terms of biology? Most fundamentally, it resides unconsciously in the brain's topographical maps of its body parts, their location in space, and their relation to nonself objects and events outside the body. The conscious "I" is another matter, though it includes a conscious recognition of the mapped body parts and relations to nonbody objects and events. Since everything the brain does is contained in patterns of nerve impulses in interacting circuits, it seems logical to think of consciousness as existing as a complex system of such patterns, likely differing from the patterns during unconsciousness.

At the conscious level, the sense of self arises from a global brain workspace and becomes explicit, giving rise to an awareness of autonomy and the impression of free will. The human brain does not create two independent persons in one brain, but rather two kinds of processes. Unconscious and conscious operations are seamlessly integrated. Both constitute a sense of "I," differing in explicitness. Both use the same neural "hardware," but in a different way. The brain knows and feels, but when conscious, it knows what it thinks and feels.

Though scientists have not bothered to pursue in depth this fundamental point, there is a major shift in functional connectivity of networks in the brain when the brain shifts from unconscious sleep to wakeful consciousness. Functional brain scans of conscious humans reveal widespread functional connections and interactions, even during mental-rest states (Smith et al., 2015). Moreover, connectivity patterns have personal specificity, correlating with differences in age, history of drug use, socioeconomic status, personality traits, and various intelligence tests. More robust connectivity was noted in subjects with positive traits, such as more education, above-average memory ability, and physical stamina, while lower connectivity levels were present in people with negative personal traits, such as aggressive behavior, smoking, alcohol abuse, or recent marijuana use. Is it not reasonable to expect that some, if not many, of these brain wiring changes are products of conscious processes and choices? Is it not possible that conscious thoughts shape mental characteristics because they shape brain connectivity?

If consciousness has agency, it can act on behalf of the brain and body to best serve its self-interests. Especially for the present purposes, we need to expand our understanding of what a "being" or "agent" is. The embodied brain is a being characterized by its enormous set of electrical circuitry with nonlinear operations that are not always predictable.

Consciousness even affects certain bodily actions indirectly. For example, consciousness does not mediate visceral and reflex control, but it most surely can modulate it. We can deliberately slow our breathing. The conscious mindfulness of meditation lowers blood pressure and slows heart rate. During times of stress we can reduce the release of adrenalin and cortisol by mentally calming ourselves down. For biological functions that we can't control directly, we can seek out medication, as with treatment for itch, muscle soreness, inhalants for asthma, and drugs for diarrhea and vomiting.

We can restrain withdrawal reflexes, as by lying on a bed of nails or keeping our hand in ice water, up to a point of course. We can choose not to orient toward stimuli that are too objectionable or irrelevant.

The agency that results from seamless interaction of the brain's "two minds" has been described by saying that the neocortical executive networks "disappear into a distributed conglomerate of potential causes of action that include environment and practically any part of the brain" (Fuster, 2013). However, I would not use the word "disappear." I think the executive networks control conscious processes and recruit neuronal resources from elsewhere in the brain. A similar stance is taken by Bernard Baars (2003), who contends that executive control operates unconsciously but that consciousness captures that neural machinery to enable new ways for multiple networks to cooperate in implementing conscious intentions.

Because consciousness is necessary, by definition, for free will to be expressed, I and others agree with Nahmias (2015), who claims that the free-will debate cannot be resolved until we explain consciousness. But I think he is wrong to assert that consciousness cannot be "reducible to, nor distinct from, the working of our brain."

3.6 FREE WON'T

Arguments for conscious agency should also include active suppression of agency, that is, whether freedom exists in the will not to act. Baumeister, Masicampo, and Vohs (2011) have argued that "free won't" is indirect, seeming to exist primarily in a conscious ability to resist unconscious impulses or select from competing options.

Some critics of free will feel forced to admit that humans have "free won't." In fact, Ben Libet himself obtained results that he interpreted to allow for the existence of free won't veto power. Most recently, veto capability has been demonstrated in a study showing that people can cancel (veto) a movement that the readiness potential indicates is already determined and under way (Schultze-Kraft et al., 2015).But "free won't" is equivalent to a "free will" to make a nonaction choice. To illustrate the point, suppose in an experiment you are directed to repeatedly indicate which of three colored circles (pink, green, blue) you prefer. If you are male, you might have an innate, programmed preference for blue, which means that you might pick blue 65% of the time. But you can decide, for whatever reason, to pick pink consistently. You would be able to pick pink 100% of the time. Have you freely vetoed blue? Or have you freely chosen to pick pink?

The ability to consciously and freely cancel a decision or stop an action is frequently glossed over. Free won't can be just as important as free will. The capacity for free won't restores the reality of personal responsibility. Even if we were unable to choose to act freely, our capacity to freely stop actions that violate conscience, social norms, or the law means we should be held accountable for failures to veto when it was appropriate to do so. Vetoes are choices or decisions. Why would our brains be free to make these kinds of negative decisions but not decisions that create positive action?

3.7 POSSIBLE MECHANISMS FOR CONSCIOUS AGENCY

Agency can be defined as the initiation of intentions, memory recall, value assessment, decisions, and planning that is orchestrated to produce overt action (Klemm, 2015). Agency is the willed expression of such actions. Of course, many human actions are generated unconsciously. But that is not proof that some actions cannot be generated consciously.

Regions in the neocortex have been identified that are associated with the feeling of consciously mediated executive control. This network operates in top-down fashion to regulate many thoughts and behaviors (Beck, 2008). The network includes the anterior cingulate cortex, dorsolateral prefrontal cortex, and parietal cortex. It may still be an open question whether such control arises freely, but it is clear that the control system operates during consciousness to mediate such functions as initiative, choice, prioritization, and planning in concert with value assessment being performed by the limbic system's reward centers in the hypothalamus, ventral tegmental area, and nucleus accumbens. Executive control also includes the veto of potential actions.

The ultimate holy grail of science may be discovering the mechanisms of conscious agency. The enigma may be unsolvable. What follows below, however, is at least a preliminary explanation of how consciousness could reduce to brain function and, moreover, gain the capacity to "do things."

Let us consider how sensory and motor realizations are represented in the brain. First, we must emphasize that sensory and motor realities are not literal realities inside the brain, but rather are abstracted and represented in the brain by patterns of nerve impulses in specified neural circuits. Such circuit impulse patterns (CIPs) are clearly the basis for unconscious processes involving sensations, feelings, and movement commands. The pattern of activity changes over time in correspondence with input and processing requirements, as illustrated in Fig. 3.1. The figure is analogous to what is seen in the time course of metabolic activity in fMRI scans.

This figure illustrates unconscious processing. However, similar neural principles could be invoked to explain consciousness. Conscious sense of self is a system property of brain that creates and processes

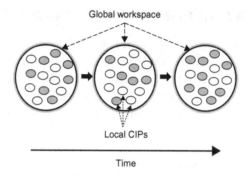

Figure 3.1 The brain represents information as distinct patterns of circuit impulse patterns (CIPs) (small circles) *within the brain's global workspace* (large outer circles). *Most of the local CIPs are probably functionally coupled* (small gray circles) *at a given moment. The dynamics of brain processing are reflected in time-varying shifts in connectivity of network activation patterns as sensory input and processing requirements change. Patterns of activated networks* (small gray circles) *change over time with the shifting attention and processing of sensations, thoughts, feelings, and movement commands.*

impulse representations, and thus the brain surely has the inherent capability to use this fundamental CIP mechanism for generating conscious states. Processing of these impulse representations no doubt occurs combinatorially over all neurons involved in the networks supporting consciousness, integrating the patterns within and among all the circuits engaged in a given process at each instant. While this description can be trivialized as just another materialistic process of nature, conscious awareness of what these patterns represent is indeed a profound property of the human brain. It takes biology to the new and unique level of conscious sense of self and nonself.

Consider the possibility that the state of consciousness emerges when a special supplementary combinatorial aggregation of CIPs is triggered from within the global workspace. Thus, the brain could develop the consciousness set of CIP patterns, while at the same time continuing its unconscious operations (Fig. 3.2). Some of these patterns become accessible to consciousness whenever there is spatio-temporal overlap in certain regions of neocortex. This is consistent with the fact that small regions of the neocortex mediate perception of specific conscious content (Koch, 2012). The consciousness CIP set can have its own dynamics wherein the set shifts selective conscious attention by spatiotemporal shifting of the overlap patterns and thus enabling a continuously variable stream of consciousness awareness and agency. This could be the basis for selective attention.

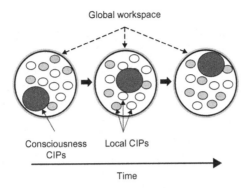

Figure 3.2 Illustration of how a special set of CIPs (large dark circles) *might constitute consciousness, which operates simultaneously with unconscious processing. Spatiotemporal shifting of the consciousness CIP set could task-dependently monitor and interact with a subset of the unconscious CIPs whenever there is overlap of shared circuitry and impulse patterns.*

The important point is the overlap of conscious and unconscious CIPs. The overlap allows bidirectional sharing of the nerve impulse representations, while at the same time giving consciousness access to the same circuitry that "does things" during unconscious processes. This could provide the neural resources that consciousness would need for agency.

How might that "sharing" work? One way is synchronous activity between and among various neuronal ensembles, which is indirectly monitored by spectral coherence analysis of extracellular voltage fields (that is, the EEG). Before more explanation, we need to understand the accepted but inappropriate term, "field potentials," used to describe extracellular voltage fields. Currents in the brain are carried by ions moving through various resistive paths, thus yielding measurable voltages. All voltages of whatever source have fields with strengths that diminish rapidly with distance from the origin. In the brain, there are two kinds of voltages, those that occur across cell membranes and those spread by the ionic current flow throughout extracellular fluid. The direct signaling in brain exists in the form of transmembrane voltage spikes, called spikes or action potentials, that propagate along the defined pathways of the fibers of neurons. The extracellular voltages spread geometrically with no such guided propagation. They just spread passively with diminishing strength (volume conduction) throughout the whole brain, which allows the spread because the water and salt nature of the extracellular environment are electrically conductive. I will coin a term here, compound extracellular voltage (CEV), to describe these voltages because they are compounded from a summation of extracellular voltages that come from multiple sources such as such sodium spike currents, calcium currents, intrinsic membrane potential oscillations, glia currents, gap—junction currents, and most importantly, excitatory and inhibitory currents in the synapses. Although CEVs lack direct signaling ability, when sufficiently large they can modulate spiking in nearby neurons, a process called ephaptic or electrotonic transmission. You can think of such effects as electrostatic, except that CEVs are constantly changing in their mixed-frequency content and amplitude.

Because CIPs are the main cause of CEVs, researchers commonly regard CEVs as a proxy indicator of neuronal signaling, with impulse firing rate correlating with the dominant frequency. Coherence analysis (defined as the ratio of cross-correlations of a frequency domain signal from two sites divided by the autocorrelation of each site) is typically determined by correlating field potentials from pairs of electrode sites in or over different brain areas. Any two coherent site pairings suggest functional coupling and cooperative processing. Temporal synchrony of underlying spike activity and synaptic voltages contribute greatly to the magnitude of CEVs in a given location and of course influences the degree of synchrony in other connected areas of the neocortex.

Synchrony reflects shared communication. One can infer from synchrony measurements (frequency coherence) that the amplitude of CEVs at any given site is proportional to the synchronous activity of nearby neurons. This effect has special relevance in the neocortex, which is the seat of consciousness.

Not only can voltage fields bias the membrane potential of dendrites and axons with the field, the field itself can propagate widely via intracortically fiber tracts at speeds of roughly 3–11 m/second in conscious humans. In the neocortex such "traveling waves" are not only seen as evoked by stimuli but also occur without external stimuli (Patten et al., 2012). The spread travels bidirectionally along intracortical fiber projections, especially cortico-cortical fibers. In the absence of stimulus, waves in the alpha and theta bands are seen to move from the front to the back of the scalp, whereas with a visual task-related stimulus, the propagation is back to front.

Traveling waves provide a mechanism for coordinating neural activity over wide expanses of cortical networks. More precisely stated, it is the CIPs that generate the traveling waves that are doing the traveling and provide the coordination.

Consciousness may well be generated by synchronization of standing waves in different cortical locations or by the propagation dynamics underlying traveling waves. A key test of this idea would be to compare these phenomena in the same subjects under such varying conditions as the unconsciousness of non-REM sleep, concussion, and anesthesia, with the consciousness states of REM sleep and alert wakefulness while various complex mental tasks are performed, to my knowledge no such studies have been done.

The structure of cortical columns is central to oscillatory coherence and traveling waves. The throughput neurons (those that receive input from elsewhere and relay processed output to other columns and subcortical structures and spinal cord) are large pyramidal-shaped cells. Pyramidal cells are unique in having two tufts of dendrites (the usual site where excitatory and inhibitory inputs summate algebraically and may summate to trigger impulse discharge near the cell body, which then propagate down to the targets of that cell's axon branches) (Spratling, 2002). One dendritic tuft occurs at the cortical surface (apical) and the other close to the cell body (basal). The important point is that each tuft receives input from different sources (Fig. 3.3).

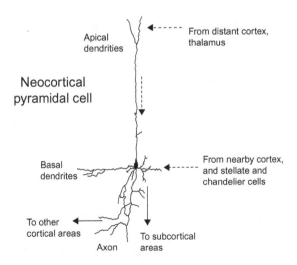

Figure 3.3 Illustration of the main neuron of the neocortex, the pyramidal cell, showing a few of its inputs and outputs. The differential input in basal and apical dendrites could provide a way for overlapped simultaneously operation of the CIPs of unconscious and conscious states.

This provides a cell-level mechanism for overlapped integration of conscious and unconscious processes.

We should emphasize the structure and connections of the pyramidal cells, because they are the basic building block of the neocortex and the only cells that send projections outside the cortex. Moreover, the pyramidal cells of humans have up to 30-fold more dendritic spines than in other parts of the neocortex (Elston et al., 2006).

This anatomical arrangement means that simultaneous inputs from other cortical areas, subcortical brain regions, and sensory inputs from cranial and spinal nerves can generate a rich mixture of different CEVs simultaneously in the same target pyramidal neuron. Indeed, coherences in hippocampal cortex are affected by where microelectrode pairs are placed with cortical columns (which are explained in chapter 5) (reviewed by Buzsáki & Schonberg, 2015). What is needed is to evaluate CEV coherence within and between neocortical pyramidal cells during shifting states of sleep and various shifting conditions during consciousness. The required technology is available.

Moreover, changes in synchrony can occur between frequencies detected at a given electrode site. For example, cross-frequency coupling has been studied in multiple zones of cortex, wherein high-frequency (gamma) voltages vary in a time-locked way with the phase of slower (theta) activity coming from that same area (Buzsáki & Schonberg, 2015). The consequence of such coupling derives from multiple input sources whose neurons may be oscillating at distinct frequencies and/or phases relative to each other. Those oscillating input streams may tune activity in the target neuron if there is a proper phase relationship.

This anatomical segregation likely has functional implications that might be expressed in differential and relatively independent contributions to CEVs in those dendritic arbors and with coherences with the soma and dendritic arbors in other locations of neocortex. Of course, the nonlinear processing in these two zones eventually combines in the cell's area of spike generation, but the inputs to different dendritic regions may contribute to this output in different ways. Whatever the functional consequence of these separate processing zones, the significance is magnified because pyramidal axons form long-range projections to other cortical and subcortical regions. Spratling (2002) even suggests that apical inputs can influence computational processing at the basal dendrites and may be considered "a source of reinforcement or supervision."

The idea of overlapped function of unconscious and conscious modes might be testable by simultaneous microelectrode recording in apical and basal zones as well as in the spike-generating region of the soma of pyramidal cells. Moreover, functional coupling could be monitored by looking for coherent relationships between the pyramidal soma spikes in one area of cortex with the CEVs of the dendritic areas of other pyramidal cells that receive those projections (Buzsáki & Schonberg, 2015). Coherence of high-frequency CEVs (>30 waves/second) appear to be crucially associated with consciousness (Singer, 1999). Perhaps such synchrony binds CIPs into the cognitive gestalt that we call consciousness.

The various signals in a given cortical locus superimpose, yielding a compounded multifrequency signal (CEV). Some of these multiple frequencies may arise from differential processing in the neurons near a given electrode site. CEVs at various vertical points along the neuronal axis seem similar but do not contain identical frequencies (Buzsáki, Anastassiou, & Koch, 2012). Thus, it is possible that these frequencies have shifting coherences within the cortical layers spanned by pyramidal neurons that a given electrode is monitoring.

3.8 THE BRAIN'S CONSCIOUS AVATAR

Murphy (2011) holds that the complex system character of brain function can become an agent, a point that I first made more explicit in my argument that the conscious mind is equivalent to an avatar that the human brain unleashes to serve the interests of body and brain (Klemm, 2011c). Just as a computer-game avatar is a being computationally generated as a proxy for the gamer, the human brain's unique computational capacities create its own kind of proxy. The conscious state is realized when the requisite CIP state is generated, and, once generated, the brain's conscious avatar serves as the brain's coagent. The brain's avatar has nonlinear functional mechanisms and adaptive flexibility could enable creative thought and a degree of freedom in arriving at choices and decisions, as will be explored further in chapter 5.

The avatar is more than just "the little man" in the sensory and motor cortex. This homunculus is a set of consciousness-unique CIPs that encompasses wide expanses of neocortical global workspace, and uses the homunculus more as a frame of reference for sensation and release of instructions for movement. Because the avatar CIPs overlap with the CIPs of unconscious processing, at least some of the unconscious representations and processes are shared and preserved within the avatar CIPs. The sharing is enabled by physical connection points linking the respective neural networks and probably enhanced by synchronization of CIP activity in distributed regions of the brain.

> *Conscious being exists as a unique set of circuit impulse patterns that have agency. The set is, in effect, the brain's avatar.*

Such a brain can program itself in response to the avatar's explicit awareness of self and situational contingencies. Thus a free-will critic might claim that such programming precludes any possibility of free will. But what about the choices a self-aware autonomous brain makes in choosing experiences that will have programming consequences? Are those choices also inevitably preprogrammed? The brain does typically have a predisposition to prefer experiences that have positive reinforcing effects for it or its body. But common experience reveals that the brain often makes mistakes in this regard. In any case, the choice occurs in the context of seeking positive reinforcement or avoiding negative reinforcement. In that sense, our freedom is definitely constrained—but not necessarily eliminated.

Unlike computer avatars, human brain avatars can generate their own choices and actions. The conscious avatar has a sense of agency, namely, that it can make decisions and do things. Free-will critics say that this sense of agency is a lie the brain tells itself. But this is a circular argument. In order to defend the notion that a conscious brain has no free will, one must assume that consciousness has no agency capability. On what grounds can such an assumption be made? In fact, numerous observations support the opposite assumption, as I will suggest in chapter "Free-Will-Dependent Human Thought and Behaviors." The avatar can also serve to assist unconsciously driven decisions. When decisions are needed, the unconscious mind can benefit from having an outside reviewer, so to speak, to provide the level of perspective, reason, and judgment that only its avatar can provide. To do that most effectively, the avatar must be allowed a degree of free will.

Human avatars have the capacity to sense, intend, evaluate, decide, plan, and direct action. A conscious avatar is self-aware of its body and much of the environment in which that body operates. It also senses a larger context of its own thought, including beliefs, wishes, decisions, plans, and the like. It can sense how it teaches its unconscious partner in terms of attitudes, emotions, understanding, capabilities, motor skills, and ideas.

But the avatar does not have awareness for everything being sensed. Conscious attention is selective and at any moment the conscious brain is blind to much of its input (Tye, 2009). But it is further true that the conscious brain can perceive sensations that do not even exist, for example by "filling in" the missing elements in ambiguous-figure perception (Klemm, Li, & Hernandez, 2000).

The key point is this: brains neurally construct conscious perception. The brain is wired for perceptual construction. An eye, for example, does not see in three dimensions—the brain constructs depth perception from the input from both eyes. What is constructed is not inevitable. If the brain has some freedom in perceptual construction, might it not also have some freedom in constructing choices and decisions?

> *The conscious avatar is the "ghost in the machine."*

There is clear evidence for the avatar notion. Neural CIPs relate indirectly to the EEG, and the EEG frequencies change as the state of consciousness changes. The EEG gets most of its signal from the part of the brain that is closest to scalp electrodes, the neocortex, the locus of consciousness. In the most alert conscious states, the EEG is characterized by low voltages at relatively high frequencies (40-plus waves/second). In relaxed, meditative states, the voltages in the EEG are larger and slower, often in the range of 8–12 waves/second. In emotionally agitated states, 4–7 waves/second appear. In unconscious states, the waveforms are especially large and still slower (from 1 to 4/second). In coma, the EEG voltages decrease to the point of extinction.

Likewise, changing CIPs in the neocortex changes conscious thought. Cerebral strokes, for example, can eliminate circuit impulse activity in the affected region, and simultaneously eliminate the thoughts normally mediated by the affected area. Similar effects can be produced by injecting anesthetic into a localized blood supply to the cortex. Naturally occurring epilepsy simultaneously changes impulse activity patterns, consciousness, and thought. Contemporary research with localized transcranial magnetic stimulation demonstrates highly specific changes in conscious thought.

The avatar, existing as a special combinatorial code of nerve impulse patterns, exists as a representation of conscious self-identity. Its CIPs represent its own self to itself, as opposed to the representations of the external world that it is aware of. Unlike the concept of dualism, the "ghost in the machine" exists as the avatar emerging from within. But it only seems like a ghost. It is made of the real substance of CIPs.

Neuroscience is not yet in the early stage of trying to identify what is unique about the impulse patterns that occur during consciousness versus those during unconsciousness. Current research relies on EEG and fMRI techniques, and these are not sufficient. Such measures are not the same as CIPs (Klemm, 2014).

For the purposes of the free-will debate, it does not matter whether we understand what is unique about the impulse patterns of consciousness. What does matter is the fact that consciousness is a self-aware, autonomous state of mind that probably has agency and a degree of free will. I would reverse Descartes' famous axiom, "I think, therefore I am," to "I am, therefore I think—and will!"

I am, therefore I think—and will!

Consciousness clearly resides in the neocortex. Destroy or otherwise inactivate the neocortex and consciousness disappears. The neocortex makes our conscious choices. But this does not by itself address the argument over whether such choices are willed freely. But it is consistent with the intuitive notion that consciousness can do things.

3.9 WHEN THE AVATAR GOES TO SLEEP

Our personhood dissolves into unconscious oblivion when we go to sleep. Yet we wake up to be the same person. That means that the CIPs of consciousness personhood can be regenerated.

In detail, however, we are not the same person after a sleep episode as before (Klemm, 2012b). Experiences during the preceding wakefulness period are consolidated in memory and may in the process modify not only the preexisting store of external knowledge but also make minor adjustments in personality, attitude, belief, and the like. Older adults cannot be the same as they were as a child or teenager. These changes are realized in consciousness, and it seems likely that consciousness had a role in making the changes to personality, attitude, beliefs, etc.

Another point: If the unconscious mind drives our choices, and behavior, why can't it do much when asleep? Agency and any free will are clearly lost in sleep. As far as we know, the main functions that operate during nondreaming sleep are memory consolidation and visceral function adjustments. Dream sleep is another matter. In dream sleep, we become a virtual agent who participates in conscious dream scenarios. Though dreams have psychological implications, the neural mechanisms that trigger dreaming likely help to trigger wakefulness (Klemm, 2011b).

One issue I have never heard or read discussed is the matter of whether the unconscious mind operates the same during nondreaming sleep as it does during wakefulness. For example, we have no evidence that the unconscious mind during nondream sleep can reason, carry on conversations, hear music, make plans, feel pain, or generate intentions. If all the many things that the unconscious mind cannot do during sleep exist during wakefulness, then there is no credence to the argument that the unconscious mind is the source of all agency during wakefulness. Those who challenge this logic need to suggest what capability wakefulness imparts to the unconscious mind. How would this happen? If the unconscious mind does have new powers during wakefulness, we should consider that it could have been enabled by coupled interaction with conscious processes.

Can one be awake and unconscious? Of course under certain abnormal conditions of disease or drugs, one can have greatly distorted consciousness. Well-established evidence suggests to me that wakefulness and consciousness are both triggered concurrently by the same brainstem arousal mechanisms acting on the neocortex (Klemm, 2014).

Unlike the unconscious mode of sleep, the conscious avatar that is launched in wakefulness can attend, intend, decide, plan, and act because it exists as self-aware CIPs. How can being awake and having the impulse patterns of the avatar reveal the meaning of its own impulse patterns? That is the crux of the "hard problem" of explaining consciousness that science seems ill-equipped to explain.

I suppose that unconscious processes could have CIPs that direct attentiveness, decision, and so on, but the patterns cannot be the same as during consciousness because two different representations are concurrent: information about what is sensed or thought, and explicit awareness of the information that is sensed or thought.

I have summarized the neuroscience of wakefulness elsewhere (Klemm, 2014). The point to make here is that wakefulness is a physiological *state* that is necessary but perhaps not sufficient for consciousness. Consciousness may also require recurrent loops of feedback from the thalamus to the neocortex. Since the classic discovery of Moruzzi and Magoun on brainstem-mediated arousal, we know that wakefulness crucially depends on activation from the brainstem reticular formation, which has a network of neurons in the central core of the brainstem that monitor spinal and cranial nerve inputs and provide activating collateral axonal projections into all parts of the neocortex. Also involved is the drive of the neocortex from acetylcholine neurotransmitter released from the nucleus basalis Meynert in the basal forebrain.

Nonmammals can be awake, but are unlikely to be conscious because they lack the neocortex that is necessary for generating consciousness. In such animals, we should not confuse behavioral quiescence with mammalian sleep.

Wakefulness is necessary but insufficient for consciousness.

Brain waves (that is, the EEG) are clear indicators of wakefulness or sleep. EEG recordings from lower animals such as fish, amphibians, and most reptiles show a wakefulness pattern, even when the animals are behaviorally quiescent (Klemm, 1973). This should not, however, be taken to mean that these primitive species are conscious. It just shows that they are awake, perhaps perpetually.

So what are the brain-wave correlates of consciousness? We don't know, but such correlates are an active area of research (Koch, 2004). However, nobody has framed the issue around CIPs. I think that key brain areas and the CIPs of consciousness could be identified by tracking them in the same person during progression of various stages of consciousness, particularly during sleep and the transition into and out of sleep.

I think the search for meaningful correlates could be expedited by monitoring the synchrony of specific field potential frequencies in multiple brain areas in various stages of sleep and wakefulness. There is good reason to think that frequency coherence is involved in consciousness. An experiment conducted in my lab demonstrated oscillatory coherence changes during the transition between unconscious and conscious realization of ambiguous-figure stimuli (Klemm et al., 2000). When looking at such figures, such as the famous vase–face illusion, a viewer has one of two perceptions (either a vase or a face, but not both at the same time). When a viewer switches from one conscious realization to another, there are major increases in coherence in multiple frequency bands among many scalp areas. Note that for a given perception, the stimulus impinging on the eye is constant. Only the conscious percept changes.

Many more recent studies strongly suggest that conscious states are accompanied, and perhaps created by, phase locking of activity from multiple distributed cortical populations. Unfortunately, very little has been done to test the hypothesis that phase locking of key brain areas will co-vary with various transition states along the consciousness continuum.

But numerous studies of a single state suggest that oscillatory synchronization is a crucial component. For example, when the conscious mind is grappling with a difficult problem, many areas of cerebral cortex generate high-frequency (gamma) brain waves. During high-intensity thought these oscillations often synchronize across multiple neuronal ensembles.

At this point, it is useful to consider those inherent properties of brain function that likely enable a degree of free will in directing thought and behavior. These properties include variance, unpredictability, patience, value judgments, language, working memory, deliberate memorization, reason, character development, will power, planning, and creativity. At the very minimum these functions reinforce the notion that consciousness is much more than a passive observer.

CHAPTER 4

Free-Will-Dependent Human Thought and Behaviors

In many kinds of new learning, the conscious mind has to teach the unconscious mind in a top-down way. Since in new learning an algorithm may not preexist unconsciously, the conscious brain must surely have some freedom in deciding how to proceed with the teaching and the learning. The conscious mind probably also initiates the use of mnemonics, deliberate practice, and recall strategies.

In contrast, deterministic decisions are driven "bottom-up" from sensory imperatives, either extant or from past learning, and are greatly influenced by inherited or learned propensities. "Top-down" decisions, however, are conscious mind operations that produce voluntary results. Thus, there is the opportunity at least for some components of top-down operations to have a degree of freedom in selection among alternative options. Moreover, top-down processes can monitor the consequences of decisions, even those generated unconsciously, and can moderate or veto actions. Several areas of neocortex contain circuits that regulate executive control over conscious decisions and actions. It is here perhaps that researchers should look for physiological indications of the degrees of freedom in circuit operations.

Of course human decisions are constrained by biology, the programming of past learning, present contingencies, and viable alternative choices. But constraint is not equivalent to deterministic preclusion of free will. As is common in anti-free-will arguments, accepting this premise forecloses further consideration of the issue.

However constrained, it is still possible for a person to violate biological or logical constraints and make a choice that is hard, unexpected, irrational, or unwise. Indeed, a person can even choose not to decide or to veto choices as the progression of options develops. In such cases, a person may even reframe the decision as a goal, one that needs to be

Making a Scientific Case for Conscious Agency and Free Will. DOI: http://dx.doi.org/10.1016/B978-0-12-805153-5.00004-3

achieved before a good decision can be made. For example, a person may decide to be a philanthropist, but must first create a goal and decide how to make enough money to be able to give it away.

When the results of subconscious processing become explicit in consciousness, they open a new window of opportunity to modify or veto unconscious preferences. Consciousness provides an opportunity to exert freely made choices as a consequence of explicit reasoned analysis of situational contingencies and expected decision consequences. This includes of course the free choice to veto certain preprogrammed options because they would not be useful or wise.

Skeptics might claim the reasoning behind such guidance is not free, but rather programmed by predilection, past experiences, and unconscious intelligence that create the analytical criteria used in reasoning. People would be assumed to make wise choices because they have unconsciously learned the criteria that lead to wise choices.

But how then do we explain why people make unwise and even destructive choices? Such choices may be random or driven by unconscious emotional imperatives and are thus not free. But sometimes what seems to be unreasonable to observers may not be so to the person making the choice.

Multiple human thoughts and behaviors are hard to explain without invoking a free-will cause. Critics will argue that these are specious and unprovable arguments. But they are hard, if not impossible, to disprove. They are also reasonable to suppose.

4.1 CAPACITIES OF THE CONSCIOUS AVATAR

One scholar suggests that possible agency-related functions of consciousness include: (1) serving as a scratch pad for selection of choices and agency, (2) long-range planning, (3) construction and storage of memories, (4) retrieval of stored unconscious memories for use and modification, and (5) "troubleshooting" and reflective analysis and decision-making (Mandler, 2003). Subsequently in this chapter, I will expand Mandler's list.

Take the case of constructing memories through the conscious deployment of mnemonic devices. In such instances, the conscious mind actively does things to improve its memory ability. Or consider how conscious participation in n-back training can train the brain to expand its working memory capacity.

But what about actual bodily actions commanded by conscious executive control? I have three reasons to suggest that consciousness can control at least some of what the body does:

1. Behaviors cannot be performed unless the brain is awake and conscious.
2. Both consciousness and executive functions share some of the same neurons and neocortical circuitry that mediate unconscious reflexes and stereotyped responses.
3. Quality of executive functions is proportional to the degree of conscious alertness, being greatly diminished as the brain lapses into inattentiveness and drowsiness.

All conscious people feel that they are free agents, even people who deny it on philosophical or scientific grounds. Why is this feeling so universal and pervasive if it is not real? When I go to a movie, I know that I am only an observer. But in the real world, my consciousness informs me that I am not limited to observing a movie but can make things happen in that world. The sense of agency is surely real. The only issue that remains is whether any of that agency is created out of conscious thinking.

Self-control occurs most obviously during consciousness. It is in consciousness that humans think that they can make choices and restrain desires that are too impulsive, irrational, or emotional.

A main argument for conscious agency is that much of what we do cannot come from the unconscious mind. Life confronts us all with novel situations for which the brain has no preexisting unconscious algorithm to cause us to respond appropriately. The conscious brain flexibly constructs algorithms for decisions, often dependent on the use of self-talk silent language and sometimes equations. Conscious awareness of movement options enables control that is not possible unconsciously, especially during skill acquisition.

Consider, for example, the brain's problem in learning complex movements like riding a bicycle or touch typing or learning golf. The unconscious mind has no way to know in advance what to do and thus cannot provide instructions. Remember when you learned to ride a bicycle? You had to think consciously about how to make your body do multiple things simultaneously: push the pedals, steer the handlebars, and shift balance. There is no way the unconscious mind could know how to coordinate these things. There is no way that the unconscious mind alone could make you succeed. Your conscious mind had to make the evaluate changing motion dynamics, make decisions, observe the results, adjust as needed and thereby teach the unconscious mind how to accomplish this task. A whole series of decisions had to be made that the unconscious mind could not make because it did not yet know how. Your conscious mind made these decisions, only constrained by gravity that made you fall down if you didn't do things right. You could fall down, and probably did at first. Later of course, once the unconscious mind was taught, the correct decisions and movements could be conducted unconsciously. Thus free will can convert to determinism.

These same principles apply to any new complex task, such as learning a foreign language or a musical instrument, where the unconscious mind does not yet know what to do. All decisions are made consciously at first and the choices are not predetermined. The conscious mind decides to do the new things well or poorly, or even quit. The point here is that in new learning the brain makes certain adjustments among multiple options that may not be compulsory.

After a new task is well learned during consciousness, it can become transferred to unconscious systems (Koch, 2012), forming what we could call a habit or automatism. Habits are likely to be driven by deterministic unconscious processes. That can be a good thing, for it is a way for the conscious mind to make the choices that program the unconscious mind to automate appropriate behaviors. Forming bad habits is likewise an option. A given choice may not inevitably be mandatory.

Conscious intentions to decide may well be unconsciously driven "bottom-up," but they can also result from top-down conscious processes even when there is no compelling need. We may decide to have steak for dinner instead of broccoli. The need to decide only comes from hunger, but what to eat is an option, though the bias will usually be for steak. Also, we can fast. What biological urge compels that?

Numerous scholars argue that it is wrong to impute the "conscious mind" as an agent because they assume that this requires unscientific dualism. They use dualism to stack the argumentation deck. Fuster, for example, considers consciousness as a state of subjective awareness and that states are incapable of function. He says that "all these processes (of choice, decision, planning, and creating) are based on preexistent information" in the brain. That is patently false. The conscious mind selectively directs attention to targets for new learning that eventually creates information for reasoning and creative thought that was not preexisting.

A conscious mind does not have to be outside the brain in order to have agency. Conscious attention and other conscious functions do not need to guide the cerebral cortex because they *are* the cerebral cortex functions of executive control. The cerebral cortex does not need a dualistic mind to guide its agency, because it is the source of the agent (that is, the avatar). When the cortex operates in conscious mode, it exists as a special set of CIPs.

When in sleep mode, our conscious being is deactivated, yet still retains a capacity for relaunch from the stored synaptic weightings of personhood. Our personhood thus exists as stored memory. The synaptic weightings are adjusted daily in terms of connection patterns among neurons, number and sizes of synapses, biochemical capacity for synthesis, storage, release, and uptake of neurotransmitters, binding properties of postsynaptic receptor molecules, and the biochemical capacity of intracellular second-messenger system cascades.

My position reduces to one of certain neural activity patterns having executive control that can act as an agent of will because it is capable of the necessary facilitation/inhibition influence over decision-making circuitry. This is the means by which the neocortical executive control circuitry functions as an autonomous avatar. This still leaves open the question of whether the avatar's control can be freely exerted. But, at a minimum, we are not forced to reject free will on the grounds that there is some sort of dualistic "conscious mind" floating around the brain that directs willed choices.

Another flaw in anti-free-will positions derives from the restrictive stimulus—response determinism that has dominated neuroscience since its inception as a discipline. Yet this conceptual myopia fails to consider the active role of consciousness in selective attention and the whole host of capabilities that cannot be fully performed unconsciously (see below).

In other words, consciousness is more than a state. It is a *being*. But this view may create the conundrum that every person is schizoid, existing as an unconscious being (as when sleeping) and a conscious being (when awake). Worse, when awake, unconscious brain functions are still quite active. So yes, in this sense we are schizoid, but the seamless way the brain handles the interface and integration of unconscious and conscious functions should cause us to think of ourselves as one embodied brain being that can perform neural functions in one of two operational modes or both, even at the same time.

Fig. 4.1 serves to illustrate how the complex thinking can benefit from interaction of conscious and unconscious thinking.

Anytime we are awake, the conscious avatar is active (on-line). When asleep, the avatar is stored in the "hard drive" of synaptic weightings in the set of circuits that contain the capacity for regenerating the impulse patterns that represent the conscious self. That self may undergo subtle changes as a result of each day's experiences, and these in turn modify the impulse patterns of self and may produce lasting changes as those patterns are stored in modified synaptic weightings.

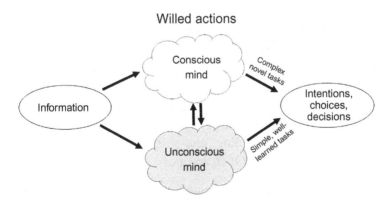

Figure 4.1 Conscious and unconscious minds interact seamlessly. Each mode of thinking adjusts for different categories of thinking and, separately or together, has agency.

The avatar might be thought of as a virtual being that is rebooted each time a sleeping brain awakens. But it is not virtual. It is real, just as patterns of electrons flowing in computer circuitry are real, materialistic processes. And like electrons in integrated circuits, nerve impulses can do things. Those impulse patterns are a code, a code that if repeated enough to alter synaptic weightings, can change the very circuits from which it is being generated. In other words, the conscious mind can change its mind.

4.2 ATTENTIVENESS AND DIRECTING ATTENTION TO SELECTIVE TARGETS

When humans are confronted with a strong, novel stimulus, they typically respond with an unconscious reflex orientation and attentiveness to the stimulus. But in many of daily life experiences, we consciously select what stimuli to attend. We can decide to pay attention and focus on certain stimuli or we can deliberately "tune out" stimuli that are unwanted. Typically, we are not compelled to selectively attend or ignore, and thus whatever choice we make could be said to be freely made. Moreover, consciousness can actually train the brain to focus, as is manifest in experienced meditators and a variety of professionals.

The sleeping unconscious mind has no such selective attention. Even in the quasiconscious dream stages of sleep, dream content is bizarre because of the greatly reduced executive control of selective attention to dream content.

4.3 SELF-CONSTRUCTED CONSCIOUS WILLFUL PURPOSES: A MATTER OF DEGREE

Can we assume, as the illusory free-will position holds, that no one is personally responsible for their intentions and subsequent acts? I think that the issue about free will is not so much whether we have any, but whether we exercise the free will we do have. Certainly, people vary widely in self-control, discipline, and will power. Are those capacities built-in and unchangeable? Of course not.

Philosopher Immanuel Kant argued that we are essentially free and fail to live up to the promise of such freedom if we allow biology and external forces to rule us. Philosopher Jean-Paul Sartre, despite the flaws I see in his irreligious and political beliefs, had it right in his existentialist arguments that humans do not have a fixed nature handed out from biology. Humans, he says, make their own nature out of the freedom to do so in the environment in which they exist. Humans are regarded, in existential philosophy, as independently acting and responsible conscious beings. Each individual brain is uniquely constructed from life experience, and I contend that much of that experience is self-selected and constructed by conscious willful choices. Science does not hold that the conscious mind is independent of genetics or environmental programming. But each mind is an independent being, acting in the world as a distinct entity, not as some Borg-like unit in a collective.

We know that people can overcome bad habits and addictions, behave in more responsible ways, and achieve great improbable things, even in the face of major obstacles. So, the issue at hand is not whether humans are personally responsible for their behavior, but whether their actions are inevitably dictated by prior programming. When we behave well, do we deserve any credit? When we act badly, did "the Devil make us do it?"

The conscious mind that believes in its power to choose must also believe that it can exert some control over the automatic processes and purposes of its unconscious mind. No doubt, the conscious mind has its own automatic purposes, but some of these are subject to veto or modification via free will. Such minds believe they can train and discipline their minds, bending them to freely willed purposes. Such minds say to themselves, "I can quit smoking." "I can make myself learn how to be an engineer." "I will make this marriage work." "I will not lie, cheat, or steal, nor tolerate those who do." And so on.

Now it is true that all of us commonly surrender our decision-making autonomy to our unconscious baser instincts and compulsions. Much of our behavior is unthinking, knee-jerk responsiveness. Our capacity for free will is limited to our willingness to claim and exert it.

People who believe that humans have no free will are hard-pressed to explain how certain of our choices and actions can be compelled. What is it that compels foolish or deviant behavior? Are we compelled to believe in God or to be an atheist? What compels us to accept one moral code over any other? Are we compelled to become a certain kind of person, with no option to "improve" in any self-determined way? Are we compelled in our choices of learning experiences? If so, what or who does the compelling? It seems unreasonable to assert that we are inevitable victims of genetics and experience or a robotic unconscious mind.

Current debates about determinism and free will tend to obscure the important matters of our humanness. Consider the argument that brains make our choices. The door to understanding what is really going on is slammed shut by assertions that value choices and the decisions that flow from them cannot be free because they are caused by neural circuit impulse patterns. Where is the evidence that there is no freedom of action in neural circuits? Such claims distract us from a proper framing of the issues about human choices and personal responsibility.

> *Our capacity for free will is limited to our willingness to claim and exert it.*

While it is true that brain circuitry is programmed by genetics and experience, that same circuitry generates a conscious mind that makes choices about who and what to interact with and what experiences to value, promote, and allow. The conscious mind can insist that some lessons of experience need to be remembered and valued, while others are not. In short, the conscious mind in the process of generating purpose gets to help shape what its unconscious mind becomes. Indeed, that is probably the main value of consciousness.

Those who dismiss conscious agency have no supporting evidence. Disparaging conscious agency is often used as a tactic to excuse bad behavior. For example, many defense lawyers increasingly use neuroscience inappropriately to convince jurors that the defendant was not responsible for the evil deeds. Lawyers even have a name for this kind of defense: "diminished capacity." Indeed, to them the whole notion of evil might seem inappropriate. Lawyers are adept at stressing mitigating circumstances where criminal behavior was caused, they say, by a terrible upbringing, poverty, social discrimination, or brain injury. To be sure, most murderers have been found to have a standard profile that includes childhood abuse and some kind of neurological or psychiatric disorder. But many nonmurderers have a similar profile. How can lack of free will explain such difference? The reality is that most people have brains that can learn social norms and choose socially appropriate behavior. Ignoring those norms is a choice.

A most disturbing book, written by Laurence Tancredi (2005), uncritically argues that human morals are "hard-wired," with the "wiring" created by genetics and molded by uncontrollable forces of life experiences. Tancredi is a lawyer and practicing psychiatrist. Not surprisingly, the poster boy for his arguments was a psychopathic serial killer, Ricky Green, who was abused as a child and had relatives with serious mental problems. Thus, Tancredi stresses that bad genes and bad treatment as a child made Green become a "biologically driven" murderer. Yet, in recounting the case history, it became clear that Green was not insane. He was fully aware of his childhood past, and was fully conscious of, even remorseful over, his murders. He was also aware that his out-of-control episodes were triggered by the combination of sex and alcohol. So, it was clear, even to Green, that his crimes could have been prevented or at least minimized by avoiding alcohol. He apparently was not an alcoholic who had no control over drinking. Even if we give the benefit of doubt to the conclusion that Green could not control himself, it is a stretch to argue that the uncontrollability of psychopaths applies to everybody else. One would have to argue that normal people are only normal because they got good genes and had a childhood in which their mental health was not damaged. Virtue, like conscious agency and free will, would be an illusion.

Interestingly, Tancredi acknowledges that the brain is changeable if skilled therapists provide structured rehabilitation for dysfunctional thinking. If the brain is changeable, and neuroscience clearly shows that it is, why does such change only occur with a skilled therapist? A common view is that dysfunctional people are victims who seemingly can't help themselves, and therefore psychiatrists and government are duty-bound to mold the brains of people so they overcome bad genes and whatever bad experiences life has thrust upon them. People, robots that they supposedly are, lack the power to nurture their own brain. Thus, government must create a cultural and educational environment in which humans are molded to conform to some predefined state of normality. Does this remind you of Aldous Huxley's *Brave New World*?

Also not considered by Tancredi and his crowd is that dysfunctional people might have become that way through their own freely determined bad choices along their life's journey. Those bad choices may have even sculpted maladaptive changes in their brain's unconscious function.

Tancredi acknowledges that many people have overcome bad genes and very traumatic childhoods. Sexually abused children do not necessarily become sexual predators as adults and may, in fact, become crusaders to protect children from abuse. But in his view they don't get any credit for a freely chosen decision to live a wholesome and helpful life. Their virtue is attributed to some undefined necessity, not to anything they voluntarily chose to do. How then do we account for the effect of schools and religious teachings? Do we conclude that it is the inner robot that decides which ideas and beliefs to accept and which to reject? If so, why do some inner robot brains accept the teachings and others reject them?

Some people believe that brain scans can sometimes seem to predict that certain people will commit crimes or other antisocial behaviors. But these are only conclusions from averaged scans among many sociopaths, and single-case predictions are not reliable. Also, nobody seems to consider the alternative to a "bad brain" cause of misbehavior: namely, that what people freely choose to do changes the brain in ways that make it bad and more likely that similar behavior will be repeated. People who voluntarily indulge mind-altering experiences, such as unsavory friends, drugs, or destructive ideologies and lifestyles, have nobody else to blame for those choices.

The evidence for brain plasticity, for good or bad, is overwhelming. Yet, this evidence tends to get ignored when excuses are sought for inappropriate behavior. It is true, of course, that children usually have little control over their circumstances and rising above it is surely hard for them, but they still have a human capacity to overcome as adults. Many millions have done just that. Denying that people have such capacity is not only wrong but dehumanizing.

Bad brains can surely cause bad behavior. But it is equally true that bad behavior can cause bad brains. What you choose to experience, think, and do sculpts brain function and anatomy to shape what you will become.

> *Bad brains can surely cause bad behavior. But it is equally true that bad behavior can cause bad brains.*

We are what we repeatedly do. As Aristotle noted:

Excellence, then, is not an act, but a habit.

The corollary is, in my view:

I will become what I repeatedly choose to do.

At this point, it should be useful to identify specific brain thought processes and behaviors during consciousness that likely involve a degree of free will. The examples provided below clearly occur when a person is awake and conscious. No evidence leads to the conclusion that these conscious thought processes and behaviors and the involved willed modifications can occur when a person is asleep. To argue that the processes I am going to enumerate do not derive from conscious free will but rather are unconscious functions, free-will deniers would also have to argue that these unconscious functions occur during sleep, but require wakefulness to enable the agency needed for expression. As you read the analysis of these processes and behaviors, ask yourself if there is any evidence that they occur during sleep.

This raises another point, almost never discussed. While we quibble over what consciousness is, perhaps we should focus on what unconsciousness is, a set of reflexes and stereotypical operations. Perhaps conscious agency is the default mode of brain operation in species that have the required neuronal architecture and physiology. Wakefulness thus could be seen as the fundamental state of vertebrates, giving rise to consciousness in species with sufficiently advanced brains.

4.4 PAIN

Pain is a perception occurring only during consciousness. Though pain has no direct agency, it is inextricably linked to action and thus is an indirect indication of the agency of consciousness. For example, when awake and conscious, a person swats at a mosquito. When asleep and unconscious, this does not happen. The perception of pain indirectly causes the conscious agency of avoidance. No one would voluntarily let a surgeon operate without any anesthesia.

4.5 FLEXIBILITY

Conscious brain function is surely flexible. Flexibility comes from inherent variance, which is a cardinal feature of brain function, especially at the circuit level of function in the neocortex where consciousness resides (Fuster, 2013). Such variance might be a prerequisite for free will and its enabling brain operations. As Fuster explains, selection among variants leads to new patterns of brain response. The issue here is whether such selection is made freely. According to Fuster there would be no possibility of a neural executor like the conscious avatar to supervise and direct the selection.

The degree of variance in the human brain relates to its nonlinear operations and the scale of capability produced by 86 billion or so neurons, each of which can have hundreds of connections, and also about ten times as many glial cells. If we take into account indirect connections, every neuron in the brain is connected to every other one. This is the ultimate in network organization. It provides infinite combinatorial power, flexibility, and capacity for dynamic change and some freedom.

4.6 UNPREDICTABILITY

Individual attitudes, beliefs, and behaviors are idiosyncratic and often unpredictable. Free-will deniers would say this diversity simply reflects the probabilistic selection of a programed unique genome or experience. But how can we really know where unexpected changes in usual attitudes, beliefs, and behavior come from? Chance or free will? Or a mixture of both?

Surely, an unexpected change in attitude or behavior could come from free choice, at least in theory. Who could predict the many Horatio Alger "rags to riches" stories? Surely, such changes are not all accidents or predetermined. Who could predict that the ignorant rag-tag boy growing up in the backwoods would become Abraham Lincoln? Who could predict that a political dirty-trick operative like Charles Colson would become leader of a nationwide prison religious ministry? Who could predict the cases where enemies reconcile? Who could predict an addict who rejects the addiction? On what grounds can we claim that free will had no role in such radical changes in behavior?

Examples like the above are neither predetermined, inevitable, nor accidental. They come because people voluntarily change, and people change because they will to do so. Nothing forced them to make such changes. These kinds of willed change often occur in spite of, not because of, biological or experiential imperatives.

4.7 PATIENCE: MAYBE MORE THAN A VIRTUE

People generally consider patience to be a virtue. Patience may also support the notion of free will. Patience is a form of veto, in that the brain defers action and vetoes an impulse to act in the present. Complex or highly important decisions call for conscious deliberation, not immediate reflex or stereotyped responses. Reflexes are indeed deterministic. But free-will deniers find themselves in the untenable position of believing that all higher human thought is little more than a set of reflexes. Suppressing reflexes or automated responses may be just the opposite of determinism (recall the finger-suppression parlor game mentioned in Chapter 1). For complex decision-making, sustained introspection and reasoning may well lead to discovery and a final decision that was not fore-ordained. How could that happen if brain networks had no freedom to arrive at their own decisions?

Even rats can show patience and veto capacity. A recent study used the same electrical marker of decision-making as used in Libet-type experiments, but studied rats in a different paradigm (Murakami, Vicente, Costa, & Mainen, 2014). Highly thirsty rats were trained to wait in place after hearing a sound cue until a second sound, a "go-for-it" sound, was heard that would give them access to water. If the rats showed the required restraint, they received a larger water reward. As the rats waited, the electrical "decision-making indicator" (a large waveform over the premotor cortex like the readiness potential in Libet's human subjects) grew in magnitude and reached a threshold at the point where rats lost patience and went for the reward. The progressive increase in neural firing is interpreted as a well-known "integrate-and-fire" mechanism, wherein activity grows until a threshold for action is reached.

But these investigators also found a second class of neurons whose firing could predict the rate at which the integrating neurons added up toward threshold. This observation of apparent preceding regulatory control enabled a new interpretation of the original Libet-type free-will experiments on illusory free will. The "integrate-and-fire" population of neurons may not be making the actual decision, but rather responding to earlier voluntary decision-making processes elsewhere that regulate the integration rate toward action threshold. This does not necessarily indicate free will, but it does suggest that rats have neurons that exert veto influence. It seems reasonable to suggest that willingness to learn, patience in this case, has some element of free will.

4.8 LANGUAGE

Consciousness is the only way humans can generate language. We also manipulate thought with language in the service of effective communication of gathering information, explaining, and persuading. Musical composition involves similar active conscious construction. In these cases, consciousness is surely "doing something," something rather amazingly unique to human species. The conscious avatar is the agent of language and music. No evidence supports the notion that the unconscious mind thinks linguistically, though I suppose it could, having been taught by conscious functions.

Do we use any free will in our choice of words and syntax? Of course, we have veto control, as in suppressing curse words or otherwise offensive comments in polite company. But positive linguistic choices occur as well. Some people have better language skills than others. To what extent is their superiority freely developed? It is certainly not automatically inevitable. On what grounds do we assert that it is not a free choice to look up a word in the dictionary and plan to use it appropriately in the future?

Language skill development is certainly aided automatically by the "osmosis" of reading widely. But skill, in language, music, or anything else, has a deliberative component of wanting to improve the skill. Deliberate organization of thought and word choice is willed. Why is not some of this done freely? We certainly have wide latitude in how language is used. What compels us to use one set of phrasing over the many other acceptable possibilities? What evidence indicates these choices are random or inevitable?

4.9 MEMORY

When situations require us to memorize certain information, the conscious mind typically initiates a strategy. This may begin with organizing information and thoughts to facilitate understanding and learning. We may consciously intend to draw conclusions and strive for insight from introspection. Based on what we learn, we are free to alter certain prior beliefs, predispositions, and attitudes.

We may willfully employ certain mnemonics and engage in deliberate practice to enhance memorization. It is true that the brain is forming memories of a day's events while asleep and obviously unconscious. But the initial encoding and working memory tasks are performed while consciousness. Moreover, conscious use of mnemonic devices profoundly enhances the formation of memory. Consider how "memory athletes" make willful selection of mnemonic tricks to do such astonishing working-memory feats as learning the sequence of a shuffled deck of cards or memorizing 80-digit number strings in a few minutes. Using these challenging mnemonics requires intense selective attention and strongly willed executive functions. You obviously cannot do such things when you are unconscious.

Skeptics will say that such feats are all driven and performed by the unconscious mind and that consciousness is just around to realize it has happened. But consciousness is also around to command retrieval and recitation of what was memorized. Try that in your sleep.

Working memory is a conscious selective attention process wherein one holds "on-line" the most currently relevant information while simultaneously blocking attention to the irrelevant. The process is under active conscious control (Baars, 2003). We know this occurs in consciousness (as when we look up a phone number), but we do not know if such processes also operate unconsciously. Since performing this operation is a task, often a difficult task, it certainly suggests that consciousness is "doing something." Where is the evidence this is an unconscious process?

Another thing we know is that working memory capacity is quite low (limited to about four totally independent items). The limitation could result from the need of the conscious brain to maintain the special way it has to orchestrate neural resources just to keep its avatar on-line.

We think with what is in the working memory. The limited capacity could be thought of as a handicap. But it may be an asset for conscious agency. Working memory operates in small chunks of information. Maybe we should think of it as informational "objects." This may be a way to supply input in small, sequential batches so that processing networks can keep up and not be overloaded. Small informational object batches can be consciously rearranged to achieve optimal sequencing for logical analysis.

No matter how you view it, the conscious mind likely makes some free choices when it engages and orchestrates working memory into our thinking. A major process of such orchestration is reasoning. When we reason, we make choices that are neither predetermined nor inevitable.

4.10 REASON

Because humans can reason consciously about the most appropriate choice among multiple options, they cannot be slaves to determinism. We all make choices we do not have to make, and reason helps make such choices. Reasoned choices often result from conscious cost/benefit analysis.

Reasoning benefits from actively recalled memory and takes time and effort. Such operations may operate unconsciously, but that is no argument that they cannot operate consciously, subject to free-won't vetoes and free-will reasoning. Indeed, because the results of conscious operations are explicit, consciousness promotes a selectivity in attention that can guide how the brain identifies, weighs, and sequences the criteria used in making a decision. What controls how selectivity is implemented? The unconscious mind does not experience the explicit sensations being registered in consciousness that the brain uses in consciously deciding what to focus on. Humans typically reason with language (or equations or music phrases). Where is the evidence that unconsciously driven decisions are based on these operations?

As mentioned earlier, a major objection to illusory free-will experiments is the implied assumption that decision-making is instanteous. But only the end of a long process appears in an instant. Decisions are inextricably bound with certain antecedent events as part of the decision-making process. Many components of antecedent events may be stored in long-term memory and retrieved into conscious working memory during the reasoning used in active decision-making. Reasoning is a primary function of consciousness. Consciousness provides a way to filter, organize, evaluate, and sequence thought in ways that can improve the quality of decisions over what unconscious processing is likely to do.

Brains evolved to serve their own best interests. Reason is a primary way these interests are served. This gives us the opportunity for some free will in making optional choices. We all know people who make bad choices and almost everyone has occasionally done so. Perhaps it is the bad choices that are not so free, but rather driven by emotions, insufficient will, and poor reasoning. Reason constrains our choices but provides the freedom to go beyond random selection of choice to make an optimal choice.

4.11 VALUE JUDGMENTS

Humans make conscious value judgments. These are surely biased by the brain's subcortical reward and aversion systems. But when conscious, the brain is aware of many of these reward and aversive contingencies and weighs the pros and cons of available options accordingly. We are consciously aware of this evaluation process. What is the biological point of being thus aware if we can't freely influence the process?

Often the weighing involves probability estimates. That is, we consciously try to predict the consequences of certain choices in terms of their likelihood of producing good or bad results. Such estimates might be made unconsciously too. But behavioral economists Kahneman and Tversky (1979) emphasize the importance of conscious value judgments, rejecting deterministic probability operations as the primary controller.

4.12 CHARACTER DEVELOPMENT

Children are not born with admirable character. Numerous studies show that young children have a natural tendency to lie, deceive, steal, and take advantage of others. They are driven by the positive reinforcement of instant gratification. They have to learn character from such sources as parents, teachers, social experiences, and religious doctrines. Do they learn this in the unconsciousness of sleep? Character learning occurs through conscious registration and introspective processing in which values are consciously chosen as valid and then programmed into the brain. The unconscious mind's innate preference for immediate gratification has to be overcome by conscious executive control, which not only delays gratification but also redefines what experiences serve as positively reinforcing.

> *Integrity and virtuous character are earned, not given.*

4.13 WILL POWER

People use will power to seek personal advancement or to overcome maladaptive attitudes and behaviors. Biological or emotional drives may prompt people to select bad choices, but in the light of conscious analysis and reason, people can and often do overcome their programmed impulses. Alternatively, people can set high expectations and goals that require great effort to accomplish. In either case, success requires active exercise of will power that can only occur when one is conscious.

How is it that a few people have extraordinary will power? If it is preordained, what are the imperatives? On what grounds do we assume that all such choices are unconsciously inevitable? Is this difference in will power explained solely by biology or even by experience? Is there any evidence? Consider the military recruit who changes from a shaggy, sloppy, and rebellious teenager into a disciplined soldier during boot camp. This person could quit, yet uses strong will power to abandon the prior stereotyped undisciplined lifetstyle. Another example is religious conversion, which is hard to explain other than as a freely chosen decision.

Humans are capable of extraordinary acts of will, even under conditions when all biological and situational imperatives are driving a contrary deterministic outcome. Consider the cocaine, opiate, or alcohol addicts who quit by sheer force of will. More common are the millions of former cigarette addicts, as I was, who will tell you most emphatically that quitting was certainly not preordained. In other words, addicts are free to quit and many do.

> *Humans are capable of extraordinary acts of will, even under conditions when all biological and situational imperatives are driving a deterministic outcome.*

Reasoning plays a key role here, where decisions are made contrary to deterministic drivers because it makes good sense to do so. Smokers may quit when they accept the likelihood that smoking may be killing them. One could argue that the will to live is a biological imperative that makes such wise choices inevitable. But then many people commit suicide, either by continuing unhealthy behaviors or by direct action. They do this in spite of the deterministic will to live.

4.14 PLANNING/FUTURE THINKING

Consciousness projects itself into the future. We believe that time is an arrow moving from past, present, and future. We envision what could or should happen in the future. We choose which steps to take to implement plans for the future. Where is the evidence that we do this in our sleep? What unconscious process compels such thinking?

Future thought is more than the Bayesian probability estimates that the brain can make by using a preexisting probability that is updated as new evidence is acquired (Knill & Pouget, 2004). Planning of actions certainly is future thinking, as are our hope, faith, desires, and goals. Such thinking surely occurs consciously, though that does not preclude unconscious contributions.

4.15 CREATIVITY

Future thinking often imagines what does not exist. Creativity may emerge. While we may get creative ideas in our dreams, dreaming is a conscious operation (Klemm, 2011b). Where is the evidence that creativity occurs only or even at all during the unconscious stages of sleep?

Moreover, creativity differs between wakefulness and dreaming. Dream content is frequently uncontrolled—there is no reality check nor directed application to problems. In conscious wakefulness, creative ideas are generated, framed, and refined to solve a problem. If the unconscious mind cannot do that in dreams and hallucinations, why should we think it has that sole capability during wakefulness creativity?

In the case of the human mind, system properties enable creativity, and creativity by definition is not predetermined. Generating a creative idea is an act of will, and I would argue comes close to being prima facie evidence for free will.

Fuster (2013) likens future thinking to an unconscious process wherein the future is imagined by past memory, dictating that "every decision, like every plan and every creative achievement, has a history behind it." Yet, how does this prove that the history predetermines the future thinking? In fact, the knowledge and understanding that reside in memory can be recursively referenced in an ongoing consciously directed process to produce a new result that was not inevitable.

How can determinism explain creativity? The combining of extant and remembered facts, ideas, and emotions in unique combinations may be necessary for creativity, but it cannot be predicted. This holds for art, music, literature, science, and even engineering. True, sometimes creativity seems to just happen by chance, but chance in this case is not random. It comes to the prepared mind, according to the famous saying of Louis Pasteur. Creative thought even in a prepared mind is not automatic. It can be actively generated in consciousness.

Humans can imagine things not yet conceived or experienced. Maybe such imagining can occur unconsciously, but surely the explicit imagining of consciousness should have some advantages. Consciousness allows for selective direction over thinking and a reality check on new ideas.

At the risk of anthropomorphism, I feel obliged to mention the creative capacities of dolphins, which are one of the few species whose brains are bigger than needed for their body size. Readers who go to Sea World or other marine parks are impressed with the dolphin shows where dolphins perform all manner of operantly conditioned acrobatics in response to human signals. Amazingly, they can deliberately choose a creative maneuver, one that they have not seen or performed before. Trainers use a cue that basically means "do something different." Dolphins not only display creative movements on command, but they can also do it in pairs with a hand signal that basically says "do it together." Upon such command, the pair swims to the bottom, tweet whistles to each other, and then leap to the surface to perform an identical movement. Either one dolphin is mimicking the other at millisecond speed or they actually discussed a plan in their chirping form of language.

I should also mention my scuba diving experience in Roatan, Honduras. There is a dolphin training camp there, and one dolphin who had apparently escaped stayed loose and free in the area. Every day as our dive boat went out to the site, this dolphin followed. He not only followed us out to the site, but rushed to the front of the boat and raced us directly underneath the bow. He rolled playfully under the bow and adjusted his speed with that of the boat. It was clear that he would not let the boat go faster than he could. After we finished the dive, we all hovered on the surface, as instructed by the dive master. The dolphin swam up to each of us in turn, saying in essence "hello" and spent a few seconds of interaction. I even touched his nose (which apparently was not appreciated). We later learned from the dive master that this dolphin does this every day with each dive trip. The point is that these behaviors seemed freely chosen, along with his living alone, which is antithetical to dolphin lifestyle. The dolphin had not been trained to do this, and there were no rewards other than an apparently intrinsic reward of interacting personally with humans.

Little of this proves free will, but it does prove that dolphins can be creative. They are not constrained to do what dolphins are biologically or "culturally" programmed to do.

When a person makes a conscious effort to generate an original thought or problem-solving solution, free will seems to be the most parsimonious explanation for the process. Creative thought involves thinking "outside the box." Does this not require some free thinking? The process includes knowing and rejecting ideas that are not creative and actively searching for a novel construction. When such intentions and processes are not mandatory, nor yield predetermined results, are they not made freely? Critics will say that the unconscious mind develops the intent to be creative and performs the cognitive processes. But where is the evidence to support that?

Creative processes entail certain consciously controlled operational principles, such as (1) emotional/cognitive state change (emotional perspectives, competition, shock factor, visual representations); (2) changing physical properties (synonym perspectives, opposite perspectives, visual abstraction, abstract associations); (3) comparison (comparing possible solutions), and (4) deconstruction/reconstruction (recombinations) Talent et al., submitted for publication). To deploy these principles, the conscious mind must question, hypothesize, imagine, organize, reason, anticipate utility, value esthetically, and respond to feedback. The avatar's ego is an intimate player. These are clearly conscious actions, not just "observations."

Consciousness is necessary to communicate creative ideas to others and to launch plans for implementation. In the light of conscious awareness, creative ideas, even if generated unconsciously, can be refined or vetoed. Are these subsequent processes inevitable?

4.16 MENTAL ILLNESS

Finally, an argument for freely willed conscious agency has to account for what happens in abnormal conscious states. The mentally ill are nonetheless conscious, though their thoughts and choices are often irrational. Do they choose to be irrational? No, of course not.

But another indicator of free will is the lack of it in mental illness. By definition, people without mental illness would have more free-will capacity than the mentally ill or even normal people with substance addictions. To deny free-will in normal people is to assert that we are all mentally ill, varying only in degree. Of course, there may be psychiatrists who think just that.

CHAPTER 5

Neuroscience May Rescue Free Will From Its Illusory Status

5.1 MARSHALING SYSTEMS-LEVEL NEUROSCIENCE

The problems of defending conscious agency and free will meet the greatest challenge when the argument depends on the notion of immaterial mind. I am not making that argument. To a scientist, mind is normally considered material, and the problem is how to explain how a materialistic brain can have any agency and freedom of action. In other words, are there nondeterministic material processes other than quantum mechanics? I think so. Neuroscience, not physics, is the appropriate scientific specialty to address the matter.

The brain's circuits electrochemically detect, process, and respond to situational needs. When several viable options are available, the brain's processing serves to weigh the pros and cons of each option to generate a choice or decision. The final choice is not assured and may not be the same if it has to be made again in the future. Does that not at least reflect some degree of freedom in this material process? More specifically, when a final choice is made consciously, is it not possible that the circuit impulse patterns (CIPs) of conscious agency have some freedom in making that choice?

Arguments over free will are often characterized by their subjective tediousness. What is needed is good neuroscience, which thus far has unfortunately been applied inappropriately to the issue. What is needed from neuroscience is understanding of the CIPs of consciousness, which I assume includes knowing the rate and interval codes being used, the circuitry that is uniquely engaged during consciousness, and the temporal synchronizations of impulse traffic across the various contributing circuit elements. To date, nobody has done much of this kind of study.

Neuroscience's most glaring inadequacy in this regard is its standard approach of reductionism. Willed behavior, free or not, comes

Making a Scientific Case for Conscious Agency and Free Will. DOI: http://dx.doi.org/10.1016/B978-0-12-805153-5.00005-5

from system-level processes, and neuroscience does not have good tools to investigate at this level.

Assumptions about determinism and illusory free will are typically based on reductionistic consideration of components of brain subsystems. But the whole of brain function is more than just the sum of its parts. Brain as a whole has different structure, properties, and functions than any one of its neurons or circuits (Uher, 2014). Particularly relevant is the enormous increase in degrees of freedom possessed by the system properties of whole brain, as opposed to any one part.

The systems-level perspective on brain function has been aptly described by Kyriazis (2015) as a global brain that "is a self-organizing system which encompasses all those humans who are connected with communication technologies, as well as the emergent properties of these connections." Its intelligence and information-processing characteristics are distributed, in contrast to that of individual components whose intelligence is localized. Its characteristics emerge from the dynamic networks and global interactions between its subsystems. In describing this relationship further, Kyriazis introduced the notion of the noeme, an emergent agent, which helps formalize the relationships involved. The noeme is a combination of a distinct physical brain function and that of an "outsourced" virtual one. What seems missing in the noeme is the element of agency, but this noeme idea does seem akin to my concept of a brain-created avatar.

The self is the creator of our actions. The brain generates the self, both unconscious and conscious. Clearly the brain has autonomous activity that leads to choice and decision. When the conscious "I" is operative, the neural activity (the avatar) might have the opportunity to decide things in a way not possible by unconscious neural processes. I suggest that the human conscious mind has significant self-organizing capability that is not totally dependent on "hard wiring," genetics, or past experience. I suggest, along with Sanguineti (2011) and others, that neural operations might generate at least some free choices and decisions.

Sanguineti (2011) rightly points out that much of human action is triggered by such things as personal activity level, special and urgent needs, social circumstance, and other factors. But the actual act decided upon is not necessarily predetermined. True, people often make bad choices from predetermined biological or experiential bias, but that does not mean they had no real choice. Such bad choices "do not deny freedom; they just show its limits." Or in the words of Einstein, "The difference between genius and stupidity is that genius has its limits." Of course, all humans have individual preferences and biases that can be expressed as probabilities. These probabilities may be voluntarily adjusted on the fly in response to environmental and situational contingencies. We often use reason and judgment to do what is not likely and refrain from what we would ordinarily do. Is this not an expression of free will?

The limits on freedom are probabilities, a point often missed in deterministic arguments. For any choice or decision, there is a range of probabilities affecting which of several options will be finally selected. For example, one might say that a drug addict has a high probability that drug use will continue and that there is not much freedom to "just say no." But there is that freedom, even though the probability is low. Saying no is a veto over the predilection imposed by the probability of saying yes. People can and do make such hard choices. The harder the choice, the more likely it would seem to require free will to make the right choice.

The neuroscience needed to pursue these issues remains to be developed. But one recent report seems to point in the right direction (Kaufman et al., 2015). This report suggests that even monkeys might have free will. Unlike much of the earlier free-will research that averaged impulse activity over many trials, this study tracked on a millisecond-by-millisecond basis what happened on each single trial. Nerve impulse activity was recorded from multiple electrodes implanted in motor and premotor cortex as monkeys decided to perform an arm movement to move a screen cursor to reach one of two targets at various locations in their visual field. Neural activity was visualized in histogram and in state-space form. The targets were placed behind visual barriers, so that movements had to be planned to go around the barriers to get to the target. Researchers could move the barriers so that trials varied from complete experimenter control in which a barrier blocked one of the choices, to moderate experimenter control wherein a barrier might be changed mid-trial to allow the monkey to change its mind, to free choices where both targets were accessible (and change of mind was possible). In trials with two viable options, the same reward was received irrespective of which target was chosen.

Monkeys did have biases toward one or the other choice. But choices varied with the relative difficulty of the two options. Barrier changes affected choice. But when both targets were accessible, monkeys made reaches to both targets. Sometimes, and usually when both targets were equally accessible, the monkeys changed their mind midstream by aborting one reach to pursue the opposite reach even though there was no compelling reason to do so.

During free choices, neural activity patterns fell within three categories. Usually, rapid and unwavering choices were produced by consistent neural activity. But in "changes of mind" choices, the neural state initially reflected one choice before changing to reflect the final choice. When behavior was indecisive, with a delay before forming a clear movement plan, so also was the neural activity indicator. In other words, during free-choice neural activity was diverse, not predetermined and would not necessarily be observable in responses averaged over many trials. This methodology is important in the study of free will because it allows tracking of the neural activity associated with changing one's mind before a final decision to commit to action is made.

The experiments documented change-of-mind behavior while at the same time showing that such changes were mediated by different patterns of nerve impulse activity. The behavioral choice was free in that both options had equal reward value and there were no constraints. So, it would seem free choice was made in the nondeterministic generation of the corresponding impulse patterns that caused the movements.

5.2 NEURAL CIRCUITRY AND THE CURRENCY OF DECISION-MAKING

One's sense of self exists as a nerve impulse pattern representation contained in interacting networks. When conscious, a presumably different set of CIPs, especially those in the neocortex, enables the brain to be aware of some of what is being represented.

The currency of any decision or choice, freely willed or not, is based on CIPs that hold and process the thought. Such CIPs include representations of sensory input, memory, constructions of decision processes, and movement commands to implement and monitor decisions.

How do CIPs contain distinct representations for each sensation, feeling, thought, and command? The generally accepted view assumes a rate code. That is, the more impulses that a neuron fires in a given moment, the more robust is the representation. But in my lab and several others, interspike interval codes have also been detected in single-neuron spike trains. That is, impulse trains from single neurons can contain embedded "bytes" or clusters of serially dependent interval patterns that have Markovian dependency levels of up to four or five and that occur with much greater incidence than predicted by chance and are erased by random shuffling of intervals (Klemm & Sherry, 1981; Sherry, Barrow, & Klemm, 1982).

More to the point of this analysis, we must ask how networks and their CIPs can have "will," as expressed consciously. There is a semantic problem here. We know that the unconscious mind makes decisions that might also be made as willed decisions in consciousness. Neural networks are largely regulated by the presence of inhibitory neurons that exist both inside and outside a given network module, like a cortical column. Inhibition is a prime mechanism for the executive control of "free won't." A possible mechanism for free will is disinhibition, which releases circuitry from inhibition and is readily demonstrated in neural networks.

If there is any freedom available to make those decisions, they reside in you (that is, in the CIPs of your avatar). But just who are you? You are not just your genes... not even your expressed genes, because expression is often controlled indirectly by your brain's choices of what to experience but even directly via hypothalamic releasing hormones that control so much of your endocrine system.

In terms of the "free-will" controversy, the "you" that matters is the conscious you. Though the interacting networks of the unconscious "you" can select one of many options, the full power of decision-making resides in the explicitness provided by consciousness, which has special capabilities outlined in chapter "Free-Will-Dependent Human Thought and Behaviors" that are not available unconsciously.

Willed behavior, free or not, does not come bottom-up from quarks to atoms to neurons to ganglia. No, willed behavior is driven top-down from the brain's system properties.

True, this is an exceedingly materialistic view of humanness. But it is the only view that science can address, and what we know, though incomplete, is well founded in observation and experiment. You may be more than your networks, but for this issue of free will, materialism needs to be the frame of reference for scientific conclusions. This is particularly relevant because free will skeptics insist on materialistic grounds.

Viewed from this perspective, the free-will issue becomes one of knowing whether neural circuit activity has any freedom of activity. Most assuredly, neural circuits, numbering in the clusters of millions organized from among the some 86 billion neurons, have incalculable degrees of freedom, in the statistical sense of the enormous number of possibilities. There is clearly an incalculable amount of flexibility in an information processing system that operates this way. By degrees of freedom, I mean that system properties enable many more output possibilities than possessed by any one subsystem and thereby the capacity for unpredictable choices or decisions. Consciousness provides a mechanism for magnifying and exploiting inherent degrees of freedom.

5.3 HOW THE BRAIN MAKES CHOICES/DECISIONS

Controversies over how freely humans make decisions are obscured by vague references to "I" or "we" making decisions. It is the brain that makes decisions. There is no doubt that many choices and decisions are made unconsciously. Colloquially, we call such decisions rash, unthinking, or even stupid. When decisions are made through reflective conscious analysis, we must confront the question of how freely the conscious brain decides or chooses among viable alternatives.

The general process of making choices and decisions involves multiple elements, some produced unconsciously, while others are more likely to be made consciously (Fig. 5.1). Notably, many of these elements are ignored by free-will deniers in their embrace of Libet-type experiments.

Decisions are network driven, and may require a continuing reverberating process within multiple interacting ensembles of neurons. Common models of decision-making commonly assume a competitive process in different circuits (reviewed by McMains and Kastner, 2011). At the neural level these processes operate as a competitive process in different pools of neurons representing each option. Eventually one pool wins out and its choice prevails. There are two main theories (1) a competitive process among the differing pools and (2) a guided-gating process.

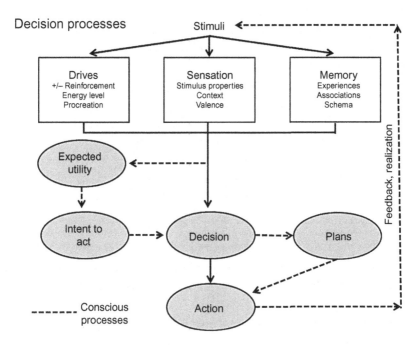

Figure 5.1 General process of decision-making. The combination of stimulus-activated drives, sensations, and memory recall generates a constellation of decision antecedents, many of which are likely to engage conscious processes (dashed lines).

These ideas emerge from study of how nonconscious processes operate. In simple reflex circuits, for example, whether or not an input generates a reflex output depends on accumulation of excitatory influence in the chain of neurons that produce the reflex. If excitation reaches the threshold for impulse discharge, then output results. It seems reasonable to apply this principle to higher brain functions that result from pools of neurons that control a given state.

Higher-level decisions may involve an accumulation process wherein noisy signals in competing circuits accumulate and are averaged (Brunton, Botvinick, & Brody, 2013) (Fig. 5.2). Averaging reduces the noise. The so-called "drift diffusion" accumulator model posits a moment-by-moment prediction of the evolving accumulation process. A key feature of the model emphasizes the importance of noise reduction.

The starting point for such accumulation may well be set by prior expectations or predictions (see later comments on chaotic initial

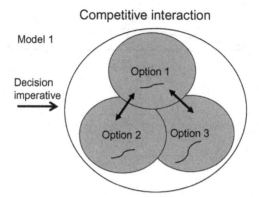

Figure 5.2 Diagram to illustrate how interacting neuronal populations could compete to reach a decision. Each alternative choice is presumed to be processed in its own local circuitry. The overlap with circuits processing other alternative choices enables sharing of impulse traffic as each population competes for dominance. Activity within a population either builds up or diminishes as interaction progresses (note the curves representing cumulative activity). At some point, a threshold is reached within one of the populations (here Option 3), making its activity dominant and representative of the final decision.

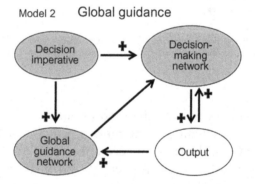

Figure 5.3 A guided-gating model for decision-making. The stimuli that create the need for a decision simultaneously activate a decision-making network and a guidance network that weights the alternative options in terms of their salience, emotional context, memory of relevant past learning, and the expected reinforcement associated with each outcome. The evaluation steps may occur in multiple circuits, but they collectively guide the decision-imperative drive to an output of one of the alternative options at the expense of others.

conditions). Moreover such initial conditions would likely affect the accumulation process and final decision (Lange, Rahnev, Donner, & Lau, 2013).

The other theory is based on a guided-gating theory (Fig. 5.3), which is similar, but here the idea is that the flow of activity in a network of brain areas steers the flow of excitation toward one option.

Input distributes in parallel to pools of decision-making neurons and is guided to regulate how much excitatory influence can accumulate in each given pool (Purcell et al., 2012). There is some evidence that the thalamic nucleus reticularis, which has extensive interactions with neocortex, is central to such guidance (Taylor & Alvai, 2003). The specific guided gating would involve inhibitory neurons that shut-down certain pathways, thus preferentially routing input to a preferred accumulating circuit.

The overlap of conscious and unconscious CIPs allows conscious-ness to influence these decision-model processes. What might that role be? In the accumulator model, conscious processes provide a new source of input that could bias activity in certain of the competing neu-ron ensembles. In guided gating, conscious input could be the agent that steers the flow of information to a population that will win out.

There is also the question of whether decision-making is bottom-up or top-down. For simple reflexes, the process is usually bottom-up. But in consciousness, there must surely be a top-down process, because con-scious decisions have to involve the neocortex, which has multiple linked executive control zones. But a recent study reveals an essential bottom-up process as well (Siegel, Buschman, & Miller, 2015). In that study, monkeys with surgically implanted electrodes in six cortical areas encompassing visual sensory pathways and executive control zones revealed a progressive surge of multiple-unit impulse discharge as input information progressed from encoding in lower brain centers to a deci-sion. The monkeys were trained to categorize a moving visual dot-pattern stimulus based on its motion, direction, or color, which they indicated by a left or right eye movement and fixation. The training incorporated a visual cue that signaled on each trial whether the monkey was to categorize on the basis of motion, direction, or color. Then the visual stimulus pattern was displayed. Unit activity was separately eval-uated for cue and stimulus. The unit activity surge in each trial revealed a dynamic progression, as if being guided, in response to the visual cue along the visual pathway flowing bottom-up into the executive control centers. Unit activity indicators of stimulus task information revealed a different dynamic. The earliest peak in activity occurred only in the infe-rior temporal cortex and V4 region of visual cortex. Then followed a feed-forward peak of activity in prefrontal cortex and lateral intraparie-tal area, followed by peaks in middle temporal area, premotor frontal eye fields, V4, and inferior temporal cortex. Sustained task information simultaneously built up in executive control prefrontal and lateral intra-parietal areas and then was broadcast as decision signals to other areas.

In this kind of paradigm, an expansive network of cortical areas is used for the combined bottom-up/top-down processing in neocortex that includes encoding the cue, memory recall of the cue-dependent task requirement, and category-assignment decision. Choice-predictive signals appeared in multiple areas before a decision was actually implemented in behavior. Studies like this provide irrefutable evidence that decisions are not momentary.

This study, as interesting and useful as it is, does not resolve the issue of free will. The monkeys in this study were, first of all, monkeys. Secondly, the choices were cued and developed through prior reward-based training. Thirdly, making the wrong choice (that is, categorization) had the negative consequence of not getting the apple-juice reward.

But the study does re-affirm a common understanding that decisions are ultimately top-down processes arising from distinct executive control regions of the neocortex that become operational during consciousness. Could this task be demonstrated in a sleeping monkey, assuming there was a nonbehavioral electrographic correlate of correct categorization? Such a study in sleeping monkeys would be problematic because there is no way to provide the apple-juice reward for correct decisions and the learned behavior would soon extinguish without reward reinforcement. You could squirt the juice reward into the mouth, but an unconscious monkey could not perceive its rewarding taste without being awakened. This makes the added point that conscious decisions are frequently affected by perception of stimuli that can only occur in consciousness. In that sense, consciousness can be essential to decision-making, even if it were not solely responsible.

With either decision-making process, decision alternatives are biased by biological relevance, past learning, current emotional state, memories, and the ongoing situation and contingencies. Multiple circuits may be involved. For example, human fMRI data indicate that the right anterior insulae contributes to the response to errors, while the anterior cingulate cortex guides implementation and moment-to-moment adjustments of cognitive control (Ham, Leff, de Boissezon, Joffe, & Sharp, 2013).

One common experimental approach is to use fMRI to identify how different brain areas change their functional connectivity, defined as linked activity changes in multiple brain areas occurring at the same time, either at rest or when performing certain tasks. For example, studies of decision-making indicate that deciding to examine the salience of a given stimulus set is accomplished by neural circuits in the dorsal anterior cingulate cortex and frontoinsular cortex, which has extensive connections with the limbic system and other subcortical areas (Zhou et al., 2010). Executive cognitive control, as might be expected for freely willed actions, is generated in this network cluster.

Remember, however, that fMRI does not measure signaling, only metabolism. The data from fMRI measures reflect radiofrequency signals mostly from water molecules and how those signals differ between two time points. From that limiting fact, scholars routinely leap to suggest a change in blood flow and capillary permeability, which then leads to the further leaps of concluding that the vascular change resulted from an oxygen shortage caused by increased neuronal electrical impulses and synaptic potentials. At the signaling level, coupling most certainly involves coherence of electrical activity, which is normally indexed by coherence of specific frequencies among brain regions.

Traumatic brain injury of white matter in dorsal anterior cingulate cortex disrupts decision-making and diminishes executive control. The study by Jilka et al. (2014) used two tasks requiring executive control, a stop-signal task (equivalent to a vetoing decision) and a movement switching task. In the stop-switch task subjects were shown left or right arrows on a screen and told to press a corresponding left or right button. But when they were shown a stop signal randomly on 20% of the trials, the subjects had to over-ride that standard instruction. In the motor-switching task subjects were to respond to blue targets on a screen with the left hand and to red targets with the right hand. But on a random 20% of tasks, subjects were told to switch the hand responses.

On both the inhibition and the switching tasks, functional connectivity increased between the executive control system described above and with other structures that are normally dominant in performing routine, monotonous tasks. Tasks were poorly performed in patients with neurodegenerative lesions in white matter linked to the anterior cingulate cortex. The normal functional coupling was not present in these patients.

What this suggests to me is that conscious decisions are enabled by dynamic changes in coherent activity within network modules. Decisions may not be made in one pool of neurons or subnetwork but rather by functional connectivity of the entire cluster of subnetworks that have ultimate executive control over decision. That decision system may well be getting its input from multiple other subnetworks that are processing the various options that are available for the decision. Such diversity of input would likely be more predominant during consciousness.

Thus, the free-will issue should be reframed in terms of the coupling of certain networks. Decisions are not made by some "ghost in the machine," nor by a black box system that sorts inputs into various predetermined output options. In other words, the brain has a conscious-avatar executive that has control over its decisions.

Regardless of how circuits make decisions, there is some evidence that network impulse pattern representations for each given option simultaneously code for expected outcome and reward value. Some models take into account reward values wherein evidence accumulation for the favored option is proportional to relative differences in reward values. In short, thoughtful decisions are made from a cost–benefit analysis. As mentioned, reward values depend on conscious perception.

These value estimates update on the fly (McCoy & Platt, 2005). Networks containing these representations accommodate such information to arrive at a decision. With either an accumulative or guided-gating process, the ultimate choice is not always self-evident or predictable. That uncertainty clearly leaves room for some freedom. In other words, the whole point of processing is to arrive at a choice or decision when automated, predetermined responses may not be biologically adaptive or optimal. The existence of a decision process, whether it be competitive accumulation or guided gating, is itself a mechanism for freedom. No one outcome is predetermined but rather it is shaped by the neural population process and its results.

Human decisions are biased by the learned consequences of prior decisions. For example, we learn through experience such things as which objects of attention are relevant and which are not. A first step in making a choice or decision is the process of attending to the options that will be considered. Is such selective attention an expression of free will? Common experience teaches that humans frequently neglect to consider all relevant possibilities in making a decision. One is certainly free to search consciously for alternative options, but that does not mean we always do that.

Recent studies have focused on what is happening in the brain as subjects voluntarily elect to discriminate a visual target or shift the focus of visual attention (Capotosto et al., 2015). Human subjects were instructed to shift attention to various spatial locations in the visual field while localized, disruptive transcranial magnetic stimulation (TMS) was delivered to scalp areas overlying different zones of the brain's visual attentiveness networks. TMS over the ventral intraparietal sulcus impaired visual target discrimination at contralateral attended locations, while TMS over the medial superior parietal lobule affected the shifting of visual attentiveness irrespective of the actual visual field. These effects were correspondingly reflected in the simultaneously recorded electroencephalogram (EEG). These findings suggest that shifting or maintaining visual attention are two different willed executive control processes that are controlled by different networks in the parietal cortex. We do not know what actually happens in these networks to determine the choice to discriminate or to shift gaze, but surely CIPs differ with the choice.

A major factor in decision-making is that choices are biased or reinforced by positive reinforcement. This should not imply that everything the brain does is based on its wiring diagram or memory of past learning. For one thing, people commonly choose what they want to learn and remember (or ignore). A role for reward expectations has recently been studied by Frank et al. (2015). In that study, fMRI and EEG monitoring revealed that the decision threshold during reinforcement learning varied as a function of trial-by-trial neural activity in the subthalamic nucleus and the medial prefrontal cortex.

The larger point, as it relates to free will, is that neural processes govern such requisite components of decision-making as selecting option targets, and setting the decision-making threshold and the reinforcement probabilities. Whether consciously derived or not, how do neural populations "decide" what to attend to and what to ignore? How is a winner-take-all threshold set? How are probabilities determined and weighted? Nobody knows.

Viewed this way, "will" becomes depersonalized. This reduces to the question of how much freedom do competing neural circuits have? In terms of statistical "degrees of freedom," the magnitude may be enormous. Even a worm's circuit of only three neurons does not always respond the same way to the same stimulus. Worms normally wander in search of attractive smells. A recent study of a common roundworm showed that the same stimulus does not always provoke the same response, suggesting that the worm makes "choices" about what to do (Gordus et al., 2015). The study identified a three-neuron circuit that controls this response, and by manipulating activity of each neuron individually and in combination, the researchers could show how each neuron contributes to the response. These neurons can have cooperative or competitive interactions that regulate final output. Neuron 1 receives the initial smell signal. Neuron 3 delivers final instructions to muscles, while neuron 2 mediates communication between 1 and 3. If all three neurons were silent, the neurons did not respond to the odor stimulus and the body did not move toward it. If all three were actively firing, the stimulus shut them off, but not always. If the first neuron alone was active when odor arrived, everything shut off. If neuron 2 or 3 were shut off individually or together, neuron 1 became more responsive.

This is a simple reflex circuit with no capacity for consciousness, but nonetheless it has a substantial degree of freedom in controlling the response to the same stimulus. Though popular press reports have called this proof of "free will," this is inappropriate for the conventional definition, which requires that the will to act occur in a conscious brain.

In conscious decision-making the role of the "I" cannot be escaped. The conscious-self "I" is the result of neural activity processes. But this "I" has sculpted its underlying circuitry through prior choices and experiences. This "I" is one facet of the brain's being which when instantiated has some autonomy and therefore is likely to have some freedom of action. Specifically, this "I" can use its reason and emotions to adjust the synaptic parameters of its decision-making circuitry in terms of setting values and probabilities.

5.4 NETWORKS IN BRAIN

Traditional analysis of human-made networks uses the concept of nodes that are connected to each other. A node is thought of as a connection or communication distribution point. Each node may be programmed to recognize, process, and distribute transmissions to other nodes.

A whole scholarly field of "neural networks" has been created by electrical engineers and computer scientists. The details of neural networks need not concern us here, but it is useful to point out some commonly used principles in their operation. Artificial neural networks often employ electronic switches arranged in stacked layers, where the output of a "sensory" input layer is fed to the next layer and that output fed to the next, and so on until a final output is reached. The output is tantamount to a choice, decision, or command.

The processing throughout the network is controlled by computer programming that sets weightings of the switches in each layer. Most artificial networks contain some form of "learning rule" that modifies the weights of the connections according to the delivered input patterns. So, for example, if the input layer is presented with a digitized image of a person's face, for example, repeated "stimulation" of the network induces error-detection feedback within its layers to adjust the weightings so that eventually the network learns to recognize that face and distinguish it from other inputs. In other words, an artificial network can perform some learning like the biological counterparts: recognize a face from seeing repeated examples of the face.

So, you might say that artificial networks can have a mind of their own. Eventually, robots may get so sophisticated that we will think of them that way. Of course, such "minds" would be unconscious, inasmuch as nobody has any inkling about how to create a conscious robot. By definition, such robots could not have free will.

But humans are not robots, and their brains do not work like artificial neural networks. Application of the ideas of artificial neural networks to the brain's neural networks seems inappropriate. I once taught a graduate-level interdisciplinary engineering course with a group of professors from electrical engineering, mathematics, and computer science. Our purpose was to teach students to understand artificial and biological neural networks. We professors became ever more convinced that the human brain cannot be duplicated in hardware and software. Brain computation is not digital, but rather analog. Brain networks cannot only change the weightings of junctions within their circuits, they can actually change the "hard-wiring" of the circuits by altering gene expression and building of new connections.

Study of real neural networks in brains is in its infancy. We can say, however, that brain network activity is flexible and readily modified by interactions among the network components. Many studies have shown that neurons in a network can dramatically change their activity levels without disturbing the overall stability of the network (reviewed by Panas et al., 2015). The study of multiunit impulse activity by the Panas group has revealed that some neurons are particularly sensitive to change, while other neurons in the network promote network stability because they are not so susceptible to change. In other words, local changes can occur without disrupting overall network stability. The flexibility afforded by this dual organization could provide a basis for information-processing freedom, commonly thought of as plasticity, even in a homeostatically regulated network.

The neural networks that matter the most for human thought and willed action are those in the cerebral outer surface, the neocortex. The cortex of human brain is called "neo" because it has a unique structure not seen in nonprimates. When viewed histologically, the human neocortex has what appear to be six layers, four of which are particularly dense with cell bodies, separated by zones dominated by fibers of those cell bodies.

When viewed in terms of functional electrical connectivity, the neocortex appears to be organized in adjacent columns, all oriented in parallel perpendicular to the surface (Mountcastle, 1998). Layer I (adjacent to the skull) contains fibers that link with other columns. Layers II and III send output to other columns, and layer III receives input from other columns. Layer IV receives input from the thalamus and also sends feedback output mainly to layers II and III of the same column. Layer VI sends output to multiple subcortical areas.

The neural architecture of a typical cortical column has been determined by stimulation and recording with microelectrodes (Fig. 5.4). Note, however, that the figure is incomplete in that each column contains numerous inhibitory neurons whose connections within the circuit have not been well-mapped. The neocortex of humans and higher primates consists of stacked arrays of adjacent layers of neurons that are wired together in feedback loops.

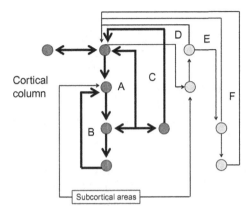

Figure 5.4 Basic functional organization of neuronal connections in a cortical column. The top of the drawing shows the tissue that lies nearest the skull. Other similar columns are stacked side by side throughout the neocortex, all oriented perpendicularly to the cortical surface. Not shown are the inhibitory neurons and their connections within the column, which are extensive. Klemm (2014), based on data and redrawn from *Annual Review of Neuroscience*, 27, 419–451.

Note that even just one cortical column contains within it at least six circuits (labeled here, A−F). A given column receives input from and distributes output to multiple other columns (in the form of nerve impulses). Thus, we see that the neocortex as a whole is a global workspace of dynamically interacting subnetworks (columns) and subsubnetworks (circuits within each column). Further support for such a view exists in the form of functional MRI findings (Schlegel et al., 2013).

Cortical columns and their complex circuitry and interconnections provide the complex system and global workspace nature of the neocortex, which yields consciousness and reasoning capacity that can sometimes freely generate intentions and decisions. The collective of cortical columns is the basis for consciousness and any conscious free decision or choice must be affected by what is happening in cortical columns.

An example of how these network interactions are involved in willed action comes from recent experiments in which spike trains were recorded from implanted electrodes during free-choice behavior in monkeys. Conscious intentions arise from interactions of at least three major neocortical areas: (1) anterior cingulate cortex, which seems to monitor and guide selective attention, (2) lateral and dorsal prefrontal areas that actually implement attentional control, and (3) ventromedial prefrontal areas that assign salience and value to attentional targets (reviewed by Oemisch et al., 2015). Impulse firing correlations between cingulate and dorsal prefrontal neurons emerged shortly after a learned selection cue and lasted for 50–200 millisecond, independently of overall firing rates. The behavioral task involved willed, voluntary choice to initiate eye movement to the appropriate cued visual target. Other nonrecorded cortical areas could have participated in the decision to change fixation by activating the prefrontal neurons, which in turn activated cingulate neurons and recurrent synchronized activity between the two areas.

Each neuron cell body operates in nonlinear fashion to inputs to summate excitatory and inhibitory inputs algebraically and thus creates a bias to allow or block transmission, depending on whether or not the discharge threshold has been reached. Moreover, the nerve impulse output may be a variable train of multiple pulses with distinct patterns of interpulse intervals. Thus, the activity in a neural circuit, unlike that of manmade circuits, has to take account of the simultaneous pattern of output of each neuron (node) in the circuit, as perhaps can be visualized in Fig. 5.5.

In such circuits neural activity is recurrent, capable of generating oscillations and staged output to other circuits. Linkage among similar circuits from staged inputs and outputs from other circuits, create the capacity for time-locked oscillations across all the linked circuits (from Klemm, 2014).

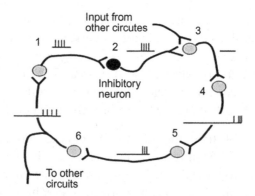

Figure 5.5 Illustration of a simple neural network. Note that each neuron may be discharging a different pattern of impulses at any given moment (see vertical bar patterns). Time delays are indicated by horizontal lines. Inhibitory neurons can pace the temporal dynamics (see the effect of neuron 3 being shut down by inhibition from neuron 2, which delays activity in neuron 4). The overarching point is that knowing the temporal dynamics of impulse discharge in any one neuron reveals very little understanding of the circuit's function. It is the collective activity of all the neurons in the circuit that matters. Unfortunately, technical limitations have prevented scientists from studying neural function from this perspective.

To extend this concept to the circuits in cortical columns, it is clear that those networks are highly flexible, capable of generating a wide array of CIPs. Moreover, output to other circuits may come from more than one neuron in a given circuit. Clearly, what is important in each cortical column is not the impulse pattern of any one neuron in the circuit but all of the neurons in each of the subcircuits within the whole column at each given moment. Combinatorial mathematics can be used to characterize the CIPs of each subcircuit, but technical constraints prevent that. The practical problem is implanting electrodes in a column that can record all of the unit activity from identified neurons at each successive moment of time and being able to identify the subcircuit that contains each neuron. This is what will be needed to discover the relevant neural correlates of consciousness and exercise of free will.

Neural correlates of consciousness and free will, whatever they are, are dynamically changing. A focus on network dynamics can lead to a radical revision of traditional thinking about the conscious sense of self. Are people ready to think that their mind is a global workspace of interacting networks?

5.5 NEURAL NETWORKS AND CHAOTIC DYNAMICS

Inability to monitor circuit-level activity of identified neurons is just one problem. The other seemingly insurmountable problem is knowing the system-level dynamics of interacting parallel pathways. Neuroscientists know little about how multiple interconnected subnetworks govern the overall function in the so-called global workspace of the whole brain. Studies have generally focused on one time series at a time, rather than the dynamics of multiple circuits operating in parallel.

Information processing in neurons depends not only on their electrophysiological properties but also on their dynamical spike-train properties. Neuronal circuits are dynamical, nonlinear systems that manifest their communication dynamics in spike trains (Izhikevich, 2007). Knowing the dynamics of spike trains and their intervals could be sufficient index of what a network is doing as it participates in information processing and messaging. As mentioned, neural circuits have emergent properties that are not reducible to the activity of any one neuron in the circuit. Thus, the spike trains of all the neurons in the network provide the true index of the circuit dynamics. Of course, full understanding requires knowing the dynamics of all the subnetworks in the global workspace. That is likely unknowable.

I know of no studies of the time series of a collection of spike trains from multiple neurons in a subcircuit. The closest we come is to evaluate the dynamics of the EEG, which can be treated as a time-series proxy of underlying impulse activity.

Is there "freedom" in a neural time series? That has to be demonstrated to support the notion of free will. But what does freedom at the circuit level mean? Is it the same as statistical "degrees of freedom?" If so, neural circuits have abundant freedom. Is it the same as mechanical degrees of freedom? Time-delayed microscopy reveals that neuronal terminals do have slow microscopic movements, which could create reversible changes in synaptic contacts. These movements seem to be essential to transport of biochemicals throughout the neural cytoplasm, but nobody seriously regards this as directly significant for signaling. But it is hardly irrelevant.

What does it mean to say that circuits can have some freedom of action? Computer chips don't seem to have any freedom. But then chips are neither alive nor wired like real neural circuits are. We need research to address this issue.

Long-accepted understanding of simple neural circuits supports the conclusion that a self-organizing network can allocate subnetwork resources in a way that could likely allow freely made choices and decisions. That is, an output selection is "freely made" when that final selection is not predetermined but rather dependent on the processing occurring within the various participating subnetworks. The choice or decision that is a result of dynamic processing is governed by the processing itself, not by inevitability. Processing that involves multiple achievable options and ultimately selects one of the options based on the results of that processing could surely be "free" in the usual meaning of the word.

If neural processing within cortical columns did not yield freely produced results, what is the purpose or value of such processing? Consider even the simple spinal reflex. Processing there usually yields a predetermined result, as in commands to muscles to produce a joint flexion or extension without any apparent element of free will. But even here, the result is not inevitable, because the simple local spinal cord circuit can be modified by its connections with adjacent spinal circuits and by both inhibitory and excitatory feedback influences that descend from several subnetworks in the brain. Thus, the coupling of local spinal circuits with supraspinal networks allows a Hindu ascetic, for example, to lie on a bed of nails, even though his local spinal circuits have a predetermined output directing the body to get off the nails. This is a poor example of how such capability is evolutionarily adaptive, but many such examples could be given that suggest that reflexes can be usefully controlled by force of will. This capacity is surely adaptive for the brain, as for example the ability to temporarily suppress stereotyped responses for immediate gratification for a delayed larger benefit.

Freely willed controls from neocortical executive control networks are exerted when the conscious brain is confronted externally with the option of agreeing to engage in a decision or to abstain. Suppose a brain decides to respond in a way contrary to the usual predilection. What changes within its participating subnetworks could produce such a decision? If the coupling is willfully changed, what would be the neural source of that will? In such a case, why couldn't the contrary will be arrived at freely? Where in the brain would that decision be made? To what extent are subnetworks "free" to change their coupling and thus the resulting decisions? Is freedom to change CIPs and voltage-field coherence equivalent to having free will? Though most patterns in a neural circuit are influenced by ongoing (and remembered) patterns in other circuits, some circuits are known to have the freedom for computationally self-organizing. To what extent are sub-networks "free" to change their coupling and thus the resulting decisions? Is freedom to change circuit-impulse patterns and voltage-field coherence equivalent to having free will? Though most patterns in a neural circuit are influenced by ongoing (and remembered) patterns in other circuits, some circuits are known to have the freedom for computationally self-organizing.

Self-organization has now been demonstrated even in robots, where a population of identical bots has built-in sensors and programming that give them self-organizing capability. Such bots can make "decisions" to do certain unpredicted things, such as marching in orderly fashion or fly around "freely" without crashing into each other (Pennisi, 2014). Human neural networks surely have more self-organizing complexity than bots.

The fact that brains can self-organize activity in their circuits should hardly be surprising. During development, self-organization of structure occurs in all tissues, and most elegantly so in the brain. This anatomical structuring is of course guided and constrained. But think about the self-organizing degrees of freedom that exist in a brain with near-infinite combinations among 86 billion neurons, each with 100−1000 connections with other neurons.

Self-organizing capability does not necessarily reflect free will. But that possibility gains standing when such dynamics occur during consciousness. Brain circuits generate different oscillating frequencies that shift over time and the frequencies vary with state of consciousness. State changes can be tracked in a crude top-down way by certain mathematical ways of characterizing the EEG, which can be taken as a proxy for the output of brain function. Such computations reveal that brain functional state changes are non-linear and contain a mixture of deterministic and non-deterministic states.

These oscillations can constrain the generation of CIPs that create the oscillations. This could be claimed as evidence for determinism. Yet, the responsiveness available in consciousness provides a way to shape CIPs both from external sensations and internal memory and reason. Specific awareness can reset the temporal dynamics of CIPs. This probably can be seen in consciously induced changes in temporal trajectories that would indicate these are not predetermined but free to the extent that conscious thought is free. However, I know of no such research.

Brain activity, however measured, looks like chaos, that is, noise. However, brain activity contains huge amounts of hidden order (otherwise the system would be dysfunctional). This order has the capacity for rapid and widespread changes, some of which might be freely generated. Conveniently, there is a nonlinear theory of dynamics, called chaos theory, that can help us understand consciousness and free will (Freeman, 2000). Chaos, a perhaps inappropriate term, is the idea that the present state of a dynamic system determines the future that is predictable only in the short term, probably milliseconds in neural systems. As data points become more widely separated in time, functional changes become erratic and less serially dependent (that is, free?) and then eventually settle into periodic temporal orbits (called attractors). A mental analogy for an attractor might be a habit or "getting in a rut." In graphical plots of chaotic trajectory, stable patterns appear as reentrant closed loops, called "attractors." At the level of thought patterns, the unpredictability provides a basis for free will. Thus viewed, each state along the unconscious–conscious continuum is a variable that links and embeds a sequence of events. The functional trajectory patterns vary with the state.

One way to display chaotic behavior is to generate a graph in which a variable at one point in time is plotted as a function of that same variable at some specified future point in time. Obviously the adjacent data points are likely to be sequentially dependent, but independence grows with distance between sampled data points. Think of it as a stock-market price today having a greater influence over tomorrow's price than it has over the price a week from now. The trajectory of successive data points in such plots may reflect underlying stability when plots show recurring closed loops or instability when plots are much less predictable. In chaotic dynamics, erratic meandering occurs until reaching tipping points (so-called "bifurcation" points) at which unexpected ordered patterns are launched. This may be the physiological basis for creative thought and free will.

In the case of the EEG, and the underlying brain function, as data points become more widely separated in time, functional changes become less serially dependent (and more free?) and then may eventually settle into periodic temporal orbits (called attractors). In graphical plots of chaotic trajectory, stable patterns appear as re-entrant closed loops, called "attractors." Fig. 5.6 illustrates these ideas by revealing the temporal dynamics from an epileptic patient during normal conscious brain function and when an epileptic seizure emerges.

As for evidence for free will, we can say that consciousness can alter its own temporal dynamics because it is itself a brain function that can keep changing the inputs to its chaotic operations. Triggering of intentional acts might arise from self-organizing chaotic states in portions of functional trajectory. The degree of free will is limited only by the limits on freedom of action within all the neural circuits in the brain's global workspace. The freshness, unpredictability, and complexity of

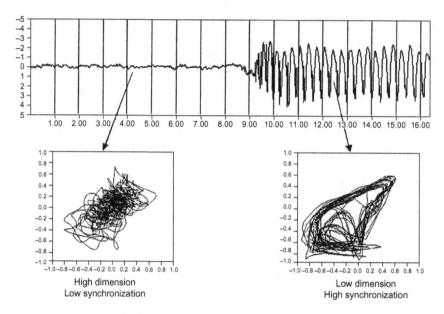

Figure 5.6 EEG trace from an epileptic patient during a normal conscious state (left part of upper diagram) and when a seizure occurs (right part of upper diagram). Corresponding phase-space plots below show that temporal dynamics are erratically unpredictable during normal consciousness but shift into a deterministic attractor trajectory during the seizure. From Stam, 2005.

our intentions and decisions are constructed in consciousness by processes that would seem to require some freedom of action within neural circuitry.

In terms of the issues involved here, this is an undeveloped field of research. A single variable, like a channel of EEG data, is often evaluated in state-space plots. But that is woefully inadequate for characterizing the dynamics of a defined neural circuit, much less the simultaneous meandering of activity in other linked circuits. Consciousness, based as it is on CIPs, is a vastly larger and more complex dynamical system than a single spike train.

Freeman, Stam, and many others claim that chaos provides a way for specific thoughts to occur without necessarily having specific causes. Consciousness intentions in particular can be self-organizing. Each module in the brain, as in cortical columns, for example, could be destabilized at any time by the influence of other modules. Extracellular voltage fields reflect these dynamics in their shifting frequencies and amplitudes. Interregional couplings are indexed by phase locking of these frequencies. To Freeman, consciousness has agency in that it constrains local fluctuations of the brain's functional modules.

Thought is a time series of multiple impulse patterns in multiple circuits. Mathematically, no time series is totally free of randomness, which in an electrical circuit is equivalent to noise. Cannot this "noise" be a source of freedom in governing the temporal chain of events, even in an unconscious brain? Spike trains can have much noise in their interval patterns. Inherent nonlinearity may amplify the noise to yield totally new dynamic properties. This process is further magnified in a system like the brain which has nonlinear feedback. Could this not contribute to freely generated creative thought during consciousness?

Intentions, choices, and decisions result from circuit self-organization in which meanings that are represented by CIPs are modified to yield new and enriched meanings and thus new intentions, choices, and decisions. The CIPs of consciousness hold back final choices and decisions until the self-organizing is sufficiently completed. Shouldn't research evaluate chaotic trajectories of brain activity along a continuum of choice or decision-making processes?

When brain function is in the dynamical "attractor" zone, it deterministically specifies the subsequent trajectory. But the dynamical meandering before evolving to that point does not seem to be deterministic. In terms of willed action, chaos can "reshape the boundaries of the system and change its trajectory" (Al-Hoorie, 2015). As he puts it, "Consciousness does not have to be an uncaused cause, but a reorganization of existing knowledge. Fate, we may argue, is not dictated by the terrain, but by whether one resists, or yields to, it. In fact, it is probably this capacity to resist attractors that makes humans unique."

The "resisting of attractors" is a reasonable speculation, but it offers no explanation for how such resistance is produced. Resetting "initial conditions" is the known way to change chaotic trajectory. Consciousness provides a way to dynamically reset the initial conditions that determine subsequent chaotic trajectories. Thus, a small change, or perturbation, of the current trajectory may lead to significantly different future behavior. Neural processing in subnetworks might provide a way to episodically reset the "initial conditions" in a chaotic system to produce an unpredicted outcome. Thus, even a small influence of consciousness can yield major consequences.

In an awake, actively engaged brain, "initial conditions" may be reset on a millisecond scale in response to external and remembered stimuli and the mental state at the time. If a conscious brain has freedom in setting "initial conditions," then even the deterministic resulting attractors are a consequence of free will. Maybe the feeling that we have certain free choices comes from an implicit recognition of underlying neural circuit dynamics that produce unexpected or creative thoughts and decisions that can reset "initial conditions." Such a process would be neither random nor predetermined. Whatever deterministic course a neural circuit may be following, the constant perturbations from the external sensory input or the processing in other circuits keep resetting the "initial conditions." The agency of conscious mind could surely produce such resetting.

Is this a different way to think about free will? Namely, the brain is free, by its feedback circuit design and complex system-level dynamics, to control itself. If a person's conscious brain (that is, the avatar) is constituted by those dynamics, and science says that is the case, does not the "person" then have at least some degree of freedom of operation? Human thought can surely perturb the system dynamics. System dynamics generate human thought, but thought alters system dynamics.

The common metaphor for chaos is that a butterfly flapping its wings in one part of the world might ultimately trigger a hurricane on the other side of the world. While chaotic systems can deterministically generate the so-called butterfly effect, this does not usually happen in the brain because the brain has inhibitory capacities (manifest most profoundly as reason) that can constrain or redirect a chaotic process. Diseases, like epilepsy or mental illness, could of course generate exceptions.

Murphy (2011) argues for the complex-system view of free will, pointing out that freedom becomes possible when a range of states becomes accessible as chaotic dynamics of brain reach the "edge of chaos," by which I assume she means bifurcation points. At the edge "the system has the freedom to explore new possibilities and may 'jump' to a new and higher form of organization."

She contends that complex adaptive systems "become causal players in their own right, partly independent of the behavior of their components, selectively influenced by the environment, and capable of pursuing their own goals." As such systems increase their complexity, as in the adult human brain, "they become more and more autonomous." They become "agents," which is exactly how I have described the set of CIPs that constitute the conscious self as an avatar.

This consideration of chaotic dynamics brings us back to the issue of network stability. A chaos-theory approach reflects network stability in the attractor zones, yet autonomous self-organizing instability during nonattractor periods. Another approach is to evaluate more short stationary time segments of data with a linear approximation model, as has recently been done in Alex Proekt's lab (Solovey et al., 2015). By "stability" they mean the ability of brain function to recover after perturbation, as from new stimuli or thought. We might suspect that during unconsciousness, as in nondreaming sleep or anesthesia, the brain is more stable in that function remains more locked into that state unless strong stimuli cause enough perturbation to cause awakening. Consciousness, however, is a state where perturbations of stimuli or thought can cause function to veer "off course" to new paths of cognitive function.

These investigators focused on a fundamental characteristic of consciousness: responsiveness. Multiple-electrode recordings from monkey neocortex during anesthesia and wakefulness provided data for autoregressive modeling showing that electrical dynamics became stable in anesthesia. The interpretation is that a conscious brain must not be rigidly locked into stability in order to be suitably responsive. It might be useful to perform this kind of analysis during consciousness where cognitive functions have to change during different levels of task complexity, decision-making, and affective state.

The brain's information processing capability lies at the heart of free will. Recall the earlier comments about Information Theory. Chaos relates directly in that the greatest amount of information in a neural network lies in the unexpected (free) activity pattern meanderings that precede bifurcations that lead to more predictable (deterministic) patterns of thought.

CHAPTER 6

Conclusion

Science thrives by falsifying hypotheses. Those hypotheses that are left standing are deemed to have some merit. Despite numerous experimental attacks, the hypothesis of free will is still standing, in spite of the fact that there appear to be far more scientists trying to attack it than there are those who want to test its validity.

Despite all the research performed on identifying the neural correlations of consciousness, very little attempt has been made to compare what is happening during unconscious states like sleep with conscious states involving "doing" various things such as registering pain, vetoing action, performing language operations, being creative, and so on.

The kind of research that could be easily done requires a comparison of field potential synchronization and traveling wave phenomena in the same subjects under such varying conditions as the unconsciousness of non-rapid eye movement (REM) sleep, concussion, and anesthesia, with the consciousness states of REM sleep and alert wakefulness while various complex mental tasks are performed.

As for free will, despite my objections to many of the claims made by Fuster (2013), I wish to conclude in much the same way as he did. He surmised (p. 192), "We are indeed free inasmuch as our brain, more specifically our cerebral cortex, has the option to take one action or another. But we are not completely free, inasmuch as its options are limited, and thus inasmuch as the brain has limits, and inasmuch as the society in which we live imposes on us limits of its own."

Ultimately, we have to think about these issues in terms of decision-making processes in those neural circuits operative during consciousness. How much freedom do competing neural circuits have? In terms of statistical "degrees of freedom," the magnitude is enormous. But more to the point, the main mechanism for free will is the existence of

Making a Scientific Case for Conscious Agency and Free Will. DOI: http://dx.doi.org/10.1016/B978-0-12-805153-5.00006-7

neural-mediated decision processes, irrespective of whether they involve competitive accumulation or guided gating. In either case, no one outcome is predetermined but rather is shaped by the neural population processing and its results. In short, decisions are not made randomly or deterministically but by the processing. When the processing is guided by conscious agency, should not the result be regarded as at least partially freely willed given the brain's self-organizing capabilities?

The analysis in this book suggests that consciousness enables the brain to do things in ways otherwise not possible. Among these is some degree of freedom in conscious agency: decision flexibility, patience, value judgments, language, working memory, active memorization, reasoned cost−benefit analysis, will power, planning, and creativity. Critics will argue that these actions could be made subconsciously, with the conscious mind being made aware of them after the fact. But the critics' view has no credible supporting evidence.

A widely accepted scientific principle known as Occam's razor should hold here. Namely, all things being more or less equal, that is, evidence for free will or no free will, we should accept the simplest explanation that makes the fewest assumptions. But things are not equal. The scientific evidence against free will is not credible, as numerous scholars have pointed out. Maybe it is a fool's errand to keep trying to prove the negative that free will does not exist.

Free will, to the extent it exists, is a top-down causal operation of conscious mind (Murphy, 2011). The concept of free will presumes consciousness and the top-down sense of "I" in which sensations, memories, and thoughts are inextricably self-referenced (Klemm, 2012a). "Will" is what the "I" decides to want or do. The point at issue is whether any aspect of will can be freely determined. Free from what? Free in the sense that a given choice or decision is not inevitably driven when achievable options are available.

How the brain makes choices and decisions will eventually receive much more scientific attention as the field of neuroscience evolves with better tools and insights. But what we know now is enough to draw some tentative conclusions. Namely, interacting and competing neural circuits contain impulse-pattern representations of the choice or decision options. Where is the "I" that makes the final conscious choice? Maybe the "I" emerges from the executive control circuit that spans cingulate, prefrontal, and parietal regions of neocortex. Consider the likelihood that the "I" is not a thing in a place but a process in a population that is the equivalent of the brain's conscious avatar.

More relevant than where the "I" is located is the *how* it can change the synaptic weightings in the decision-processing circuits that ultimately "win out" to favor a given choice. If the "I" is equivalent of a brain's avatar that knows what it knows, knows its values, feels what it feels, and is consciously aware of the predicted outcomes of available options, then the avatar's use of executive control circuit processing can bias the synaptic weightings of the processing circuit that contains the impulse-pattern representation of the optimal choice. Our true free will, to the extent it does occur, is exercised by our brain's conscious-being avatar.

Degrees of freedom, flexibility, and unpredictability, as such, are not equivalent to free will. But when network functions are amalgamated into a self-organizing, conscious being—an avatar—then that being's network capabilities can be recruited and mobilized in the service of free choices and decisions. In other words, you and I exist mentally as autonomous conscious beings that are expressions of our respective neural networks. Some sections of our networks clearly have executive power. Just because we cannot yet explain how these executive control subnetworks gain supervisory and command properties in the conscious state does not mean they do not have some degree of freedom in selecting among alternative options. Indeed, because executive control subnetworks have degrees of freedom and flexibility, they have the underlying mechanisms that can allow some degree of free choice.

Thinking about free will requires us to ponder a more fundamental issue. Who are we, in terms of our personhood? We are what has been programmed in tissue by our genetics and experiences, accidental, imposed, and chosen—including whatever portion has been freely chosen.

This personhood resides in tissue in two forms: *stored* in the form of biochemical synaptic weightings of neuronal networks and *deployed* in the form of dynamic circuit impulse patterns. The deployed version of self is malleable to new experiential programming—again including that which has been freely chosen.

Thus, humans are inescapably responsible for their choices and decisions. Even much of the neural processing underlying unconsciously driven choices was programmed by earlier consciously willed choices that created predilections and habits that biased future choices and decisions.

Brains are programmed by their experience. But free will provides another opportunity for programming in that some brain processing can consciously select and modify reactions to experience. We can override biased and stereotyped unconscious decision-making when appropriate. Moreover, we can choose many of our experiences and avoid others and thus affect our own programming. Our conscious avatar helped to sculpt what we are and continues daily influence to sculpt what we will become. Whatever we have become, we had some freedom in making it so. Likewise, we have some freedom to sculpt our future nature.

As succinctly as I know how to summarize, the crucial reasoning presented herein includes:

1. Research purporting illusory free will was inadequately designed and inappropriately interpreted.
2. Humans have a profound sense of having free will. Accordingly, they hold themselves and others accountable. Such belief is necessary for social order, legal constraints on behavior, and most religious belief systems. But identifying bad consequences of an absence of free will does not provide evidence that it exists.
3. Many neural functions and behavior are difficult to explain as the sole result of unconscious brain operations.
4. Consciousness is more than a state. It is a being, the functional equivalent of an avatar acting on behalf of embodied brain with agency and thus the potential for instantiating freely willed thoughts, choices, and behavior.
5. The human brain has enormous degrees of freedom that enable a corresponding degree of flexibility and unpredictability—even creativity.
6. When information is processed consciously, the brain has access to mechanisms for generating willed actions that are neither predetermined nor predictable.
7. Brain functions have self-organizing nonlinear dynamics that are readily reset and adjusted by situational contingencies and conscious

thought (involving language and imagery), judgment, reason, and creativity.

8. Trying to prove a negative, that there is no free will, is a fool's errand. More promising research would aim at discovering ways that a material brain might generate free will. This book has identified specific approaches.

My final thought on the issues of conscious agency and free will is that scholars need to be more open to possibilities they have rejected. Scientists, philosophers, and theologians are prone to commit what Hayek has called the "pretense of knowledge" (Hayek, 1974). He admonished us all with statements like this: "The recognition of the insuperable limits to his knowledge ought indeed to teach the student of society a lesson of humility." The problem is that scholars, perhaps more so than anybody else, are prone to enshrine their beliefs as a sure knowledge that does not actually exist.

Hasn't the time arrived when we should generate the hypothesis that humans do have a meaningful degree of free will and devise controlled experiments to test that hypothesis? We will have to look to neuroscience for enlightenment on this subject—no, not the neuroscience of the 1980s but the neuroscience of neural networks that is yet to come.

BIBLIOGRAPHY

Al-Hoorie, A. H. (2015). Human agency: Does the beach ball have free will? In Z. Dörnyei, et al. (Eds.), *Motivational dynamics in language learning* (pp. 55−72). Bristol, England: Multilingual Matters.

Baars, B. J. (1997). *In the theater of consciousness: The workspace of the mind*. New York, NY: Oxford University Press.

Baars, B. J. (2003). Working memory requires conscious processes, not vice versa. In N. Osaka (Ed.), *Neural basis of consciousness* (pp. 12−26). Philadelphia, PA: John Benjamins.

Baker, K. S., Mattingley, J. B., Chambers, C. D., & Cunnington, R. (2011). Attention and the readiness for action. *Neuropsychologia, 49*, 3303−3313. Available from: http://dx.doi.org/10.1016/j.neuropsychologia.2011.08.003.

Balaguer, M. (2010). *Free will as an open scientific problem*. Cambridge, MA: MIT Press.

Baumeister, R. F., Masicampo, E. J., & Vohs, K. D. (2011). Do conscious thoughts cause behavior? *Annual Review of Psychology, 62*(1), 331−361.

Beck, A. T. (2008). The evolution of the cognitive model of depression and its neurobiological correlates. *The American Journal of Psychiatry, 165*, 969−977.

Brunton, B. W., Botvinick, M. M., & Brody, C. D. (2013). Rats and humans can optimally accumulate evidence for decision-making. *Science, 340*, 95−98.

Buzsáki, G., Anastassiou, C. A., & Koch, C. (2012). The origin of extracellular fields and currents-EEG, ECoG, LFP and spikes. *Nature Reviews Neuroscience, 13*, 407−420.

Buzsáki, G., & Schomberg, E. W. (2015). What does gamma coherence tell us about inter-regional neural communication? *Nature Neuroscience, 18*(4), 484−489. Available from: http://dx.doi.org/10.1038/nn.3952.

Capotosto, P., et al. (2015). Dynamics of EEG rhythms support distinct visual selection mechanisms in parietal cortex: A simultaneous transcranial magnetic stimulation and EEG study. *The Journal of Neuroscience, 35*(2), 721−730.

Conel, J. L. (1963). *The postnatal development of the human cerebral cortex*. Cambridge, MA: Harvard University Press.

Dennett, D. (2014). *Reflections on free will*. <http://www.naturalism.org/Dennett_reflections_on_Harris%27s_Free_Will.pdf>.

Dennett, D. (2015). *Stop telling people they don't have free will*. <http://bigthink.com/videos/daniel-dennett-on-the-nefarious-neurosurgeon> Accessed October 18.

Di Pisapia, N. (2013). Unconscious information processing in executive control. *Frontiers in Human Neuroscience*, January 31. Available from: http://dx.doi.org/10.3389/fnhum.2013.00021.

Ellis, G. F. R. (2007). *Physics in the real universe: Time and space-time*. New York, NY: Springer.

Elston, G. N., et al. (2006). Specializations of the granular prefrontal cortex of primates: Implications for cognitive processing. *The Anatomical Record Part A, 288A*, 26−35.

Finn, E. S., et al. (2015). Functional connectome fingerprinting: Identifying individuals using patterns of brain connectivity. *Nature Neuroscience*, October 12. Available from: http://dx.doi.org/10.1038/nn.4135.

Frank, M. J., et al. (2015). fMRI and EEG predictors of dynamic decision parameters during human reinforcement learning. *The Journal of Neuroscience, 35*(2), 485−494.

Freeman, W. J. (2000). *How brains make up their minds*. New York, NY: Columbia University Press.

Fuster, J. (2013). *The neuroscience of freedom and creativity*. New York, NY: Cambridge University Press.

Gazzaniga, M. S. (1998). *The Mind's Past*. Berkeley: University of California.

Gordus, A., et al. (2015). Feedback from network states generates variability in a probabilistic olfactory circuit. *Cell*. Available from: http://dx.doi.org/10.1016/j.cell.2015.02.018.

Grill-Spector, K., & Kanwisher, N. (2005). Visual recognition. As soon as you know it is there, you know what it is. *Psychological Science, 16*(2), 152−160.

Gruber, C. W., et al. (Eds.). (2015). Constraints of agency. Explorations of theory in everyday life. *Annals of theoretical psychology* (Vol. 12). New York, NY: Springer.

Guggisberg, A. D., & Mottaz, A. (2013). Timing and awareness of movement decisions: Does consciousness really come too late? *Frontiers in Human Neuroscience, 7*, 385. Available from: http://dx.doi.org/10.3389/fnhum.2013.00385.

Haggard, P., Clark, S., & Kalogeras, J. (2002). Voluntary action and conscious awareness. *Nature Neuroscience, 5*, 382−385.

Haggard, P., & Eimer, M. (1999). On the relation between brain potentials and the awareness of voluntary movements. *Experimental Brain Research, 126*(1), 128−133.

Ham, T., Leff, A., de Boissezon, X., Joffe, A., & Sharp, D. J. (2013). Cognitive control and the salience network: An investigation of error processing and effective connectivity. *The Journal of Neuroscience, 33*(16), 7091−7098.

Hammeroff, S. R., & Woolf, N. J. (2003). Quantum consciousness. A cortical neural circuit. In N. Osaka (Ed.), *Neural basis of consciousness* (pp. 167−200). Philadelphia, PA: John Benjamins.

Hayek, F. A. (1974). *The pretence of knowledge*. Nobel Prize Lecture. <http://www.nobelprize.org/nobel_prizes/economic-sciences/laureates/1974/hayek-lecture.html>.

Herrmann, C. S., et al. (2008). Analysis of a choice-reaction task yields a new interpretation of Libet's experiments. *International Journal of Psychophysiology, 67*, 151−157.

Izhikevich, E. M. (2007). *Dynamical systems in neuroscience*. Cambridge, MA: M.I.T. Press.

Jilka, S. R., et al. (2014). Damage to the salience network and interactions with the default mode network. *The Journal of Neuroscience, 34*, 10798−10807.

Jo, H.-G., et al. (2014a). First-person approaches in neuroscience of consciousness: Brain dynamics correlate with the intention to act. *Consciousness and Cognition, 26*, 105−116.

Jo, H.-G., et al. (2014b). The readiness potential reflects intentional binding. *Frontiers in Human Neuroscience*, June 10, 2014. Available from: http://dx.doi.org/10.3389/fnhum.2014.00421.

Kahneman, D., & Tversky, A. (1979). Prospect theory: An analysis of decision under risk. *Econometrica, 47*, 263−291.

Kaufman, M. T., et al. (2015). Vacillation, indecision and hesitation in moment-by-moment decoding of monkey motor cortex. *eLifeSciences*, May 5. Available from: http://dx.doi.org/10.7554/eLife.04677.

Klein, S. (2002). Libet's temporal anomalies: A reassessment of the data. *Consciousness and Cognition, 11*, 198−214.

Klemm, W. R. (1973). Typical electroencephalograms: Vertebrates. In P. L. Altman, & D. S. Dittmer (Eds.), *Biology data book* (Vol. II, 2nd ed., pp. 254−260). Bethesda, MD: Federation of American Societies for Experimental Biology.

Klemm, W. R. (2010). Free will debates: Simple experiments are not so simple. *Advances in Cognitive Psychology, 6*(6), 47−65.

Klemm, W. R. (2011a). Neural representations of the sense of self. *Advances in Cognitive Psychology, 7,* 16−30. <http://www.ncbi.nlm.nih.gov/pmc/articles/PMC3163487/>. Available from: http://dx.doi.org/10.2478/v10053-008-0084-2.

Klemm, W. R. (2011b). Why does REM sleep occur? A wake-up hypothesis. *Frontiers in Neuroscience, 5*(73), 1−12. Available from: http://dx.doi.org/10.3389/fnsys.2011.00073.

Klemm, W. R. (2011c). *Atoms of mind. The "Ghost in the Machine" materializes.* New York, NY: Springer.

Klemm, W. R. (2012a). Sense of self and consciousness: Nature, origins, mechanisms, and implications. In A. E. Cavanna, & A. Nani (Eds.), *Consciousness: States, mechanisms and disorders.* Hauppauge, NY: Nova Science. Open access available from: <https://www.novapublishers.com/catalog/product_info.php?products_id=38801>.

Klemm, W. R. (2012b). *Memory power 101.* New York, NY: Skyhorse.

Klemm, W. R. (2013). *Core ideas in neuroscience* (2nd ed.). Bryan, TX: Benecton Press.

Klemm, W. R. (2014). *Mental biology. The new science of how brain and mind relate.* New York, NY: Prometheus.

Klemm, W. R. (2015). Neurobiology perspectives on agency: 10 axioms and 10 proposition, Chapter 4. Constraints of agency. Explorations of theory in everyday life. In G. W. Gruber, et al. (Eds.), *Annals of theoretical psychology* (Vol. 12, pp. 51−88). New York: Springer.

Klemm, W. R., Li, T. H., & Hernandez, J. L. (2000). Coherent EEG indicators of cognitive binding during ambiguous figure tasks. *Consciousness and Cognition, 9,* 66−85.

Klemm, W. R., & Sherry, C. J. (1981). Entropy measures of signal in the presence of noise: Evidence for "byte" vs. "bit" processing in the nervous system. *Experientia, 37,* 55−58.

Knill, D., & Pouget, A. (2004). The Bayesian brain: The role of uncertainty in neural coding and computation. *Trends in Neurosciences, 27*(12), 712−719.

Koch, C. (2004). *The quest for consciousness.* Englewood, CO: Roberts and Company Publishers.

Koch, C. (2012). *Consciousness. Confessions of a romantic reductionist.* Cambridge, MA: M.I.T. Press.

Kyriazis, M. (2015). Systems neuroscience in focus: From the human brain to the global brain. *Frontiers in Neuroscience.* Available from: http://dx.doi.org/10.3389/fnsys.2015.00007.

Lages, M., & Jaworska, K. (2012). How predictable are "spontaneous decisions" and "hidden intentions?" Comparing classification results based on previous responses with multivariate pattern analysis of fMRI BOLD signals. *Frontiers in Psychology,* March 6. Available from: http://dx.doi.org/10.3389/fpsyg.2012.00056.

Lange, F. P., Rahnev, D. A., Donner, T. H., & Lau, H. (2013). Pre-stimulus oscillatory activity over motor cortex reflects perceptual expectations. *The Journal of Neuroscience, 33*(4), 1400−1410. Available from: http://dx.doi.org/10.1523/JNEUROSCI.1094-12.2013.

Mandler, G. (2003). Consciousness: Respectable, useful, and probably necessary. In B. J. Baars, et al. (Eds.), *Essential sources in the scientific study of consciousness* (pp. 15−33). Cambridge, MA: M.I.T. Press.

McFadden, J., & Al-Khalili, J. (2015). *Life on the edge.* New York, NY: Crown.

McMains, S., & Kastner, S. (2011). Interactions of top-down and bottom-up mechanisms in human visual cortex. *The Journal of Neuroscience, 3*(12), 587−597.

Mele, A. (2014). *Free: Why science hasn't disproved free will.* New York, NY: Oxford.

Mountcastle, V. (1998). *Perceptual neuroscience: The cerebral cortex*. Cambridge, MA: Harvard University Press.

Murakami, M., Vicente, M. I., Costa, G. M., & Mainen, Z. F. (2014). Neural antecedents of self-initiated actions in secondary motor cortex. *Nature Neuroscience, 17*(11), 1574–1582, PMID: 25262496

Murphy, N. (2011). Avoiding neurobiological reductions in the role of downward causation in complex systems. In J. J. Sanguinetti, et al. (Eds.),*Moral behavior and free will: A neurobiological and philosophical approach* (pp. 201–222). Italy: IF Press.

Musser, G. (2015). Is the cosmos random? *Scientific American, 313*(3), 88–93.

Nahmias, E. (2011). *Is neuroscience the death of free will?* New York Times. November 13. <http://opinionator.blogs.nytimes.com/2011/11/13/is-neuroscience-the-death-of-free-will/?_php=true&_type=blogs&_r=0>.

Nahmias, E. (2015). Why we have free will. *Scientific American, January*, 77–79.

Obhi, S. S., & Haggard, P. (2004). Free will and free won't. *American Scientist, 92*, 358–365.

Oemisch, M., et al. (2015). Interareal spike-train correlations of anterior cingulate and dorsal prefrontal cortex during attention shifts. *The Journal of Neuroscience, 35*(38), 13076–13089.

Panas, D., et al. (2015). Sloppiness in spontaneously active neuronal networks. *The Journal of Neuroscience, 35*(22), 8480–8492.

Patten, T. M., Rennie, C. J., Robinson, P. A., et al. (2012). Human cortical traveling waves: dynamical properties and correlations with responses. *PLoS ONE, 7*(6), e38392. Available from: http://dx.doi.org/10.1371/journal.pone.0038392.

Penfield, W. (1978). *The mystery of the mind: A critical study of consciousness and the human brain*. Princeton, NJ: Princeton University Press.

Pennisi, E. (2014). Cooperative "bots" don't need a boss. *Science, 346*, 1444.

Pockett, S., Banks, W. P., & Gallagher, S. (Eds.), (2006). *Does consciousness cause behavior?* Cambridge, MA: The MIT Press.

Polkinghorne, J. (2009). *Theology in the context of science*. New Haven, CT: Yale University Press.

Purcell, B. A., Heitz, R. P., Cohen, J. Y., et al. (2012). Neurally constrained modeling of perceptual decision making. *Psychological Review, 117*(4), 1113–1143.

Radder, H., & Meynen, G. (2012). Does the brain "initiate" freely willed processes? A philosophy of science critique of Libet-type experiments and their interpretation. *Theory & Psychology, 23*(1), 2–21. Available from: http://dx.doi.org/10.1177/0959354312460926.

Restak, R. M. (2012). *Mind. The big questions*. London: Quercus Editions.

Rigoni, D., & Brass, M. (2014). From intentions to neurons: Social and neural consequences of disbelieving in free will. *Topoi, 33*(1), 5–12.

Sanguineti, J. J. (2011). Can free decisions be both intentional and neural operations? In J. J. Sanguinetti, et al. (Eds.),*Moral behavior and free will: A neurobiological and philosophical approach* (pp. 149–168). Italy: IF Press.

Schlegel, A., et al. (2013). Network structure and dynamics of the mental workspace. *Proceedings of the National Academy of Science*. Available from: http://dx.doi.org/10.1073/pnas.1311149110.

Schlosser, M. E. (2014). The neuroscientific study of free will. A diagnosis of the controversy. *Synthese, 191*, 245–262.

Schultze-Kraft, M., Birman, D., Rusconi, M., et al. (2015). Point of no return in vetoing self-initiated movements. *Proceedings of the National Academy of Sciences of the USA*. Available from: http://dx.doi.org/10.1073/pnas.1513569112.

Schurger, A., et al. (2012). An accumulator model for spontaneous neural activity prior to self-initiated movement. *Proceedings of the National Academy of Sciences of the United States of America*, *109*(42), E2904–E2913.

Schwab, et al. (2009). Nonlinear analysis and modeling of cortical activation and deactivation patterns in the immature fetal electrocorticogram. *Chaos. An Interdisciplinary Journal of Nonlinear Science*, *19*(1), 015111. Available from: http://dx.doi.org/10.1063/1.3100546.

Searle, J. (2007). *Freedom and neurobiology. Reflections on free will, language, and political power.* New York, NY: Columbia University Press.

Sherry, C. J., Barrow, D. L., & Klemm, W. R. (1982). Serial dependencies and Markov processes of neuronal interspike intervals from rat cerebellum. *Brain Research Bulletin*, *8*(163), 169.

Siegel, M., Buschman, T. J., & Miller, E. K. (2015). Cortical information flow during flexible sensorimotor decisions. *Science*, *348*(6241), 1352–1355.

Singer, W. (1999). Neuronal synchrony: A versatile code for the definition of relations? *Neuron*, *24*, 111–125.

Smith, S. M., et al. (2015). A positive-negative mode of population covariation links brain connectivity, demographics, and behavior. *Nature Neuroscience*, June12. Available from: http://dx.doi.org/10.1038/nn.4125.

Solovey, Y., et al. (2015). Loss of consciousness is associated with stabilization of cortical activity. *The Journal of Neuroscience*, *35*(30), 10866–10877.

Spratling, M. W. (2002). Cortical region interactions and the functional role of apical dendrites. *Behavioral and Cognitive Neuroscience Reviews*, *1*(3), 219–228.

Stam, C. J. (2005). Nonlinear dynamical analysis of EEG and MEG: Review of an emerging field. *Clin. Neurophysiol.*, *116*, 2266–2301.

Talent, W., et al. Creativity tools categorized by principle of operation: A framework for practical application. *Creativity Research*, submitted for publication.

Tancredi, L. (2005). *Hardwired behavior. What neuroscience reveals about morality.* New York, NY: Cambridge University Press.

Taylor, J. G., & Alavi, F. N. (2003). A global competitive network for attention. In B. J. Baars, et al. (Eds.),*Essential sources in the scientific study of consciousness* (pp. 1035–1058). Cambridge, MA: M.I.T. Press.

Tempia, F. (2011). Free will, perceived time and neural correlates of conscious human decisions. In J. J. Sanguineti, et al. (Eds.), *Moral behavior and free will: A neurological and philosophical approach. STOQ project* (pp. 173–185). Morolo, Italy: IF Press.

Trevena, J., & Miller, J. (2010). Brain preparation before voluntary action: Evidence against unconscious movement initiation. *Consciousness and Cognition*, *19*, 447–456.

Tye, M. (2009). *Consciousness revisited. Materialism without phenomenal concepts.* Cambridge, MA: MIT Press.

Uher, J. (2014). Agency enabled by the psyche: Exploration using the transdisciplinary philosophy-of-science paradigm for research on individuals. In C. W. Gruber, et al. (Eds.), *Constraints of agency* (pp. 177–228). New York, NY: Springer.

Wegner, D. M. (2002). *The illusion of conscious will.* Cambridge, MA: MIT Press.

Wegner, D. M. (2005). Who is the controller of controlled processes? In R. Hassin, et al. (Eds.), *The new unconscious.* Oxford: Oxford University Press.

Zhou, J., et al. (2010). Divergent network connectivity changes in behavioural variant frontotemporal dementia and Alzheimer's disease. *Brain*, *133*, 1352–1367.

INDEX

Note: Page numbers followed by "*f*" refer to figures.

Printed in the United States
By Bookmasters